Regulation of Urban Water Services

Regulation of Urban Water Services

An Overview

Enrique Cabrera and Enrique Cabrera Jr.

Published by IWA Publishing
Republic - Export Building
1 Clove Crescent
London E14 2BA, UK
Telephone: +44 (0)20 7654 5500
Email: publications@iwap.co.uk
Web: www.iwaponline.com

First published 2017
© 2017 IWA Publishing

British Library Cataloguing in Publication Data
A CIP catalogue record for this book is available from the British Library

ISBN: 9781780408170 (Hardback)
ISBN: 9781789065183 (Paperback)
ISBN: 9781780408187 (eBook)

Contents

Chapter 3
Water regulation in the UK and its relevance to Spain **39**
Michael Rouse

Chapter 4
Water regulation in Australia **65**
Andrew Speers

Chapter 5
Regulation of water services in Denmark – a utility manager's perspective   **93**
Jens M. Prisum

Chapter 6
Experiences and conclusions about regulation in Latin America and the Caribbean   **103**
Andrei Jouravlev

Chapter 7
The german experience with a self-organised water sector – key factors for an alternative to regulation
Wolf Merkel and Nicole Annett Müller

Chapter 8
Assessment of water services from the perspective of multilateral organisations. The aquarating experience
Matthias Krause and María del Rosario Navia

Chapter 9
Can a regulator contribute to resolving the main problems of the urban water cycle in Spain?
Enrique Cabrera

Chapter 10
Reasons to give grounds for urban water regulation in Spain *181*
Félix Parra

Chapter 11
International panel round table conclusions *189*
Enrique Cabrera, Enrique Cabrera Jr., and
Enrique Hernández Moreno

Foreword

The present book deals with the situation of the regulation of urban water services in the world, but also in Spain. The book originated in a workshop held in Valencia in April 2014 sponsored by Aqualia, a meeting in which the Spanish water industry and the administration decided to review and discuss the solutions that have been applied in different parts of the world for the regulation of water services. The contents of those presentations, expanded and reviewed, constitute the backbone of this volume.

Water supply services, as we know them today, date back to the 19th century, although they were consolidated in the first half of the 20th century. Wastewater treatment began in the middle of the 20th century, coinciding with the period of industrialisation following WWII. This was the logical response to growing contamination and pollution stemming from the enormous increase in urban and industrial consumption of water, although many urban drainage systems had also been built previously. Enormous investments over many decades have led to a before and after for humanity in terms of quality of life and health. This was recognised by the United States Academy of Engineering, which qualified full development of these services as the fourth most important engineering achievement of the 20th century, only behind electricity grids, the automobile and aviation. It is higher on the list than other major feats of progress (up to 16 more), as formidable as the radio and television, computers and telephony.

The rapid, disorganised implementation of these urban water systems was received the world over with enormous jubilation. But over the years the need to subject them to some kind of control has been brought to light, with good judgement, leading it to an ultimate objective: meeting the needs of citizens in a sustainable manner over time, with quality adequate for the 21st century and at the lowest possible cost. The need to renew these systems, the high quality standards they are required to meet, their intrinsic monopolist nature, their growing complexity, their social importance, and basically the demand to sustain them over time (environmentally, economically and socially) are very persuasive arguments that have advised implementing regulations in order to set the sector in order.

In Spain, the costly infrastructures inherent to the urban water cycle have been designed, managed and maintained without any clear directives and without a common denominator. It is therefore hardly surprising that the involved sectors state the need for clear, transparent rules of play. The administration too, through logic applied from the precaution stemming from the complexity involved in altering the fragile, complex water balance, admits the need to regulate these services. And it was precisely this shared concern that finally led to the workshop that was held at the end of April 2014. The workshop had two objectives; firstly, to get to know experiences in other pioneer countries and, secondly, to analyse the state of affairs in Spain from the point of view of the involved sectors.

Regulation is a topic of increasing importance for the water sector around the world. A few months after the Valencia Workshop was held, IWA published the Lisbon Charter, a key document to guide public policy and regulation. The lessons presented in this book, provided by some of the most relevant international experts on the topic, are not only relevant or applicable to countries without a clear regulatory framework, but also provide food for thought and can be used to facilitate a debate aimed to improve existing regulatory schemes.

Finally, we would like to extend our gratitude to all those that made this volume possible. Aqualia, for sponsoring the event, all the authors for relating in a very constructive manner the experiences in their countries, the Spanish administration for supporting the initiative and also IWA Publishing for the publication of this book.

The editors

Summary of the contributions

1. ENRIQUE CABRERA JR. (SPAIN)

The first chapter explains the reasons why regulating the urban water service is advisable, whilst also explaining just what regulation consists of. The reasons are obvious. Firstly, the need to guarantee the universal human right to water. Secondly, because it is a natural monopoly, strongly linked to the territory where it is practically impossible for competitors to exist simultaneously. This is because although the outsourcing processes are in some way subject to market laws, with tenders resolved, direct competition has ended even though the regulator has mechanisms to encourage it, although indirectly, as the text explains. And this is one of the tasks of regulators. Last, but buy by no means least, because providing this service becomes more complex as time goes by, and can be carried out in different ways, it is the regulators' task to guarantee that two of the principal aspects (quality and sustainability) meet the standards of demand of the 21st century. The importance of this service for citizens therefore demands some kind of supervision or control. It is therefore somewhat surprising that water in Spain is still pending regulation when there are other, non-vital services, that are regulated. Obviously the answer has to be found in history and culture.

The two principal sides of regulation are explained in this chapter (technical and economic), whilst justifying that they are both dependent on each other. Techniques today permit controlling the quality of a service, mainly through performance indicators, which is a many-sided concept. On the other hand, there is economic regulation which, supported by the principle of cost recovery, means that efficiency should be emphasised, with a guarantee of sustainability at reasonable rates in fair correlation to the quality of the provided service. Finally, and in order to better explain what is understood by regulation, the chapter discusses aspects of technical regulation.

2. JAIME MELO BAPTISTA (PORTUGAL)

This article centres much more on just what regulation is and the characteristics that must be met regarding the evolution of urban water policies in Portugal over time. This evolution in England and Australia is discussed by the two following authors. However, the article does not provide details about the characteristics of the Portuguese water industry. The article defines the most suitable regulation framework to achieve the established objectives on the basis of the principles that should govern a good water policy, particularly the urban water cycle. Among these principles the most relevant ones are the need to commit to policies that are long-term oriented, a clear institutional and legislative framework suitable for the sector, guaranteeing the three axes of sustainability (economic, environmental and social), promotion of efficiency and competence, and participation by users to guarantee transparency.

The second fundamental idea on which this paper is based is that establishing a specific regulation framework cannot be improvised. The baseline or starting point needs to be well defined (general political framework, number of companies providing services, citizen culture, etc.) and after a thorough analysis of this baseline and the objectives to be achieved, to then decide on the most suitable model to be employed as a template to specify the regulatory functions. Jaime Melo Baptista has been through this entire experience first-hand in Portugal. Many of these experiences are detailed in some of the references cited in the paper.

Finally, the author clearly describes the two plans of any regulatory process. The first of them establishes the rules of play according to which the activity is carried out in the sector, which is a task that can be directly performed by the regulator (as in the case of Portugal) or by certain technical associations (Germany). The second of them is assessment, control and follow-up to see how obligations are being fulfilled, particularly technically and economically, by the different companies. In the second plan the pioneer experience is in England (although it has been somewhat relaxed, see the article by Michael Rouse) whereas in Denmark (see the article by Prisum) the plan centres on economic aspects. Finally, the article emphasises a differentiating aspect of Portuguese regulation. There the aim is to provide the sector with technical tools (numerous technical manuals about the principal aspects of urban water management have already been published) to help the companies accomplish their objectives.

3. MICHAEL ROUSE (UNITED KINGDOM)

In this article the author describes the regulatory process in the United Kingdom in its historical context, which is a process that he states must not be linked to privatisation, which in the end has only happened in England. Wales and Scotland, although they have followed different paths, have ignored this. Regulation has been successful because it permitted modernising of and putting order in the sector, precisely because implementation of regulation was tackled when the sector was ready to deal with it. Costs started to be recovered in 1974 (implementation was considerably less popular than the regulatory process itself) and the number of companies is small, which guarantees an economy of scale. During the period between 1990 and 2012 the regulation entailed an investment of 100,000 million pounds sterling, which was absorbed by increasing bills by 37% (in real terms). It is estimated that companies financed 70% of this investment through efficiency gains. Today, the average cost structure indicates that 58.4% is for amortisation costs and therefore if infrastructures are subsidised (and supposing the same levels of efficiency are maintained) the tariffs will be reduced by more than half.

Regulation is based on three different organisations. Water quality control is the responsibility of DWI (Drinking Water Inspectorate), economic control (tariffs, investments, efficiency) is assigned to Ofwat and environmental control (of water at source and the quality of effluent) is one of the functions of the EA (Environmental Agency). All these organisations, whose tasks are described in the article, depend on one Ministry. Universal access to water is guaranteed through social tariffs (in force under the provisions of the 2010 Act) and the author points out that water bills must never exceed 5% of the available revenue. The work insists on decoupling the concept of regulation in the management method, public or private, since the regulation objectives are to guarantee sustainability of the service, promote efficiency and defend the interests of users. In other words, nothing to do with the eternal public/private debate.

The article ends by outlining the possible future of the sector in the United Kingdom, and does so by analysing the 2014 Water Bill, which fundamentally seeks to encourage a higher degree of competitiveness in the industry, which the author has concerns about. Finally, some specific principles are put forward that should govern new regulation processes, such as in Spain, in a more or less immediate future.

4. ANDREW SPEARS (AUSTRALIA)

This article provides a detailed account about the reforms in the water policy in Australia over the last three decades. Interestingly enough, the transformation that has taken place, because of the special circumstances, is very different to the case of the United Kingdom. Australia is a federation of six states and two territories, all of which are coordinated by a Federal Government, which significantly conditions its policies. The party responsible for setting objectives and directives is the Federal Government, and afterwards the States, each in its own way, adopt the specific measures to achieve them. With this political framework, in the early eighties it was decided to thoroughly reform urban water management (problems similar to those in Spain today needed to be resolved). And in order to do so, they devised two major supporting pillars.

The first of them, the general reform framework, was agreed by the Council of Australian Governments in 1994. The objectives to be accomplished were to incorporate service providing companies with a sold professional base. It was finally decided to set up public companies, but with the same operational principles as private companies and always encouraging economies of scale. These objectives include recovering all the service costs (but always protecting the most vulnerable classes), implementing consumption sensitive tariffs, separating responsibilities (the manager cannot be the party to exercise self-control), and basically to encourage competition by comparison. These objectives, it must be said, would be subscribed to by any country today. In order to encourage the different governments to introduce the necessary reforms in their respective states (particularly complete recovery of costs), the national government offered additional incentives to the ones who implemented them quickly.

The second major pillar was the National Water Initiative, signed a decade later (in 2004) with the major objective of facilitating territorial cooperation. It states that water management must be articulated on the basis of natural territorial frontiers and should not be conditioned by political frontiers. To enable this cooperation, the National Water Commission was established in charge of developing and supervising the hydrological plans in the different water basins. Among others this commission was assigned the responsibility of controlling the usage of water, ensuring the established tariffs recovered all the costs and basically regulating the water markets. Control of the companies was therefore the competence of the National Water Initiative, which became very active in encouraging water saving. This led to establishing the water labelling and saving programme that banned installation of devices (such as toilets) that did not use water efficiently. Nevertheless, and as usually happens, sanitation control was assigned to another institution, the National Health Council. This work however, does not provide details about the reforms implemented in the different states, some of whom did decide on the figure of a conventional regulator.

5. JENS M. PRISUM (DENMARK)

Experience in Denmark concerning urban water regulation is complementary to three precedents, thus giving another, highly illustrating, alternative point of view. But there is always a shared goal: to improve efficiency and competitiveness in the industry. This sector in Denmark, although entirely public, is atomised as in Spain and Germany; perhaps even more so. There are around 2500 companies, who provide services to 5 million inhabitants, all initially linked to the town halls. There is a significant difference however. The urban water service is among the most expensive in the world, with prices (several times those of Spain) that came about from economic regulation, valid since 2009, commissioned to the Danish Competition Authority. Within this framework, and seeking improved efficiency, all the municipal companies were converted into limited companies (even the smaller ones). This took place whilst retaining the public status of the companies, and consequently they cannot make profits. In other words, they can even trade on the stock market but with no profits they do not attract investors.

The analysis of Danish regulation, although from the point of view of a director of one of the main companies in the country, is very critical. The only objective of this regulation, with a "top-down" procedure, is economic control of costs and tariffs on the basis of an analysis of three items: operating expenses (OPEX), capital expenses (CAPEX) and the complementary expenses (taxes, environmental costs and those required to implement specific levels of service). The author's criticism is based on the complexity of calculations made by economists for economists which, because of their complexity, are not comprehensible for small/medium size companies (regulation is applied to services with minimum annual turnover of 200,000 m^3, i.e., those that provide services to populations equal to or over 4000 inhabitants). However, regulation does not deal with any of the technical aspects, which is why in this aspect it is a compared regulation. It is the companies, grouped around DANVA (Danish Water and Wastewater Association), who in a top-down process implemented it 10 years before.

The work concludes with a number of basic principles, which in view of the experience in Denmark, should be considered in all regulatory processes. These include the assessment period (a minimum of three years), a process of dialogue governed by simplicity, common sense and transparency.

6. ANDREI JOURAVLEV (LATIN AMERICA)

Talking about regulations in a geographical area as large as Latin America and the Caribbean can only be done with a very general outlook. Each country is different, and therefore centring on the specific details of one case makes no sense at all. This paper, as could not be otherwise, gives a general picture of the problems in the industry and analyses the role of regulation, valid in 70% of the region, and a fundamental strategy to improve it. Regulations differ vastly, since, depending on the country, there is a different action framework (state, regional and even local).

When reviewing the sector, the author analyses, although only briefly, the ups and downs over the last three decades. These include the need for private investment to upgrade the infrastructures, the problems from tariff increases to redeem them, conflicts and the role of regulators as arbitrators in many of the disagreements. These matters has been discussed in depth, even at international meetings such as the one held under the name of "International Investment Agreements, Infrastructure Investment Sustainability and Regulatory and Contractual Measures" in Lima in 2009.

Within this article, the fundamental ideas defended are the same as those previously discussed by other authors, starting off with the separation of functions, basically divided into three. The three functions can be defined as: legislation and ordinance of the sector, economic regulation and supervision of the companies, and basically a proper assessment of the services provided. Another idea the author strongly defends is the importance of efficiency as a basic strategy to guarantee the sustainability of the sector, regardless of public or private management, which according to the author is of little importance. The absolute need for regulation, whatever the nature of the service provider, and the role it must carry out to rationalise (encouraging the economy of scale) and improve the efficiency of the service are two objectives that go far beyond whether or not the providing company is public or private.

7. WOLF MERKEL AND NICOLE ANNETT MÜLLER (GERMANY)

The German water industry is a different case altogether from those described previously. There is considerable interest in Germany because all types of management coexist (public – private management), it is highly fragmented with big and small companies (a total of 6211 companies), and wide regional autonomy (specific legislation in different states). There is no regulatory body but there are formidable regulations. Technical matters are common all over the country, whereas the economic and administrative aspects depend on each region. In short, the rules of play are well established without a regulatory body to closely monitor compliance. But obviously, German culture and notable environmental education favours regular compliance with the established rules of play. Whichever the case, if somebody reports or informs of deviations (particularly in economic and tariff aspects) there are bodies with competence to penalise potential deviations. This means that the industry works exceptionally well.

The text describes the situation of the German water industry in considerable detail. The description is followed by a critical analysis about the challenges the industry faces and other aspects with a considerable margin for improvement. The strength of the sector is mostly based on excellent technical and socio-political regulations governing it. The technical side is sustained by two powerful groups, the DVGW (German Technical and Scientific Association for Gas and Water), which acting as a body promotes and establishes technical rules concerning distribution and drainage, and the DWA (German Association for Water, Wastewater and Waste) which establishes the standards of quality of water supplied and the treatment levels of effluent. The tariffs and economic aspects depend on the regional governments. Under the principle of recovering costs through tariffs (subsidies are practically non-existent), there are clear rules for calculating the tariffs and established channels for citizens to submit their complaints.

The industry, despite its importance (80 million inhabitants, 530,000 km of supply network and 540,000 km of drainage network, with significant percentages of unitary, separate and mixed networks) and its diversity, is very homogeneous in terms of quality levels, which are generally high. But it does face some challenges. The biggest being the economic aspect. With the average tariffs being four times higher than in Spain (1.91 €/m^3 for drinking water and 2.28 €/m^3 for drainage and treatment, and even higher to an average value of 2.93 €/m^3 if the tariff includes payment of rainwater), the principal challenge is to maintain the quality of the service with a stable population (tending to fall slowly) which consumes less and less water, which basically translates as a considerable reduction in revenue. This is a concern because the current scenario of climate change demands investments to be maintained, which is incompatible with the decreasing income. In addition to this, the tariff structure has a fairly unrepresentative standing charge, 10% of the income, which means that the major consumers have to cover growing burdens. Companies obviously want to increase the standing charge in order to maintain their income and share out charges more evenly. But clearly an explanation must be given to the citizens as to why they have to pay more when they are using less water. In short, Germany is a self-regulated country with the industry enjoying enviable health. But the German experience is difficult to export to other countries. No other country has the same conditions as its starting point, or the German culture.

8. MATTHIAS KRAUSE AND MARÍA DEL ROSARIO NAVIA (INTER-AMERICAN DEVELOPMENT BANK, IDB)

The final contribution by the panel of international experts is by a multilateral organisation, namely the Inter-American Development Bank (IDB). The IDB "*Within the geographical scope of its competence, it has the mission of institutionalising* transparency *in management, fostering* accountability *and consolidating mechanisms for prevention and control of* corruption, *building capability, generating and disseminating relevant knowledge and affording technical and financial support*". Hence it promotes projects such as AquaRating, which is discussed in this article. The relationship with urban water regulatory services, although indirect, is obvious. Its objective is to assess performance of water and drainage companies in a way that is both independent and certified by an external assessor, which is in line with the second plan of company control regulation referred to in the text by Jaime Melo Baptista.

AquaRating has established eight areas of company assessment (service quality, business management efficiency, financial sustainability, corporate governance, planning and execution efficiency, operating efficiency, access to the service and environmental sustainability) which do not necessarily need to coincide with what a country wants to control through regulation. As has been seen in the previous contributions, during implementation of the regulatory process the first requirement is to identify the goals that are to be achieved through regulation and, on the basis of these goals, define what is to be controlled and specify how this is to be done. The IDB's goals, by financing an industry as strategic as urban water, are to guarantee proper use of its economic funds and basically to provide the sector with technological support. These are the objectives that give this initiative meaning. Despite this, the AquaRating assessment does not necessarily match that of a regulator. But the thorough, reliable system that is built can serve as support and inspiration for any other regulatory initiative such as, for benchmarking and even for voluntary self-regulation.

9. ENRIQUE CABRERA (SPAIN)

In this contribution the main challenges and problems that water management needs to deal with in Spain are reviewed, a large number of which have been discussed previously. Among the challenges to face are the ageing infrastructure, increasing contamination, population migration from the countryside to the city, growing urbanisation, and also climate change. Among the problems, the lack of training of the decision-makers at the three levels is mentioned (political, managerial and professional), as well as atomisation of responsibilities, lack of rules of play and quality standards, price policies and lack of transparency.

The article also mentions the bases for urban water policies suitable for these new times, also discussed previously. Finally, and after two collateral reflections (the eternal argument "public or private service", and the consequences that could stem from failing to plan the transition from the current subsidised system to another to recover all costs), the author fosters regulation as the most efficient method to successfully tackle the challenges and resolve a large number of the problems involved in the urban water industry today.

10. FÉLIX PARRA (AQUALIA, SPAIN)

This author starts off by bringing attention to the current decentralisation of the Spanish water market, with thousands of regulators, in fact as many as there are municipalities, and the lack of regulatory stability required to confer this activity its legal and financial security. The lack of clear rules of play in the sector is being mitigated with the drafting of guideline documents (prepared by AEAS, the Spanish Association of Water and Wastewater Services Suppliers) and the Spanish Federation of Municipalities and Provinces, FEMP), defined in the Service Regulations and in the preparation of tariffs, guide documents that are not enforceable. Consequently, the Spanish water market today is missing a regulator to establish the necessary ordinance. The requested regulation must materialise out of dialogue between all involved parties, which will be both complex and difficult, although if it is put forward in a constructive manner and with willingness, the best solution is sure to be defined.

The chapter specifies the rules on which regulation should be based (concerning economic and technical aspects), the competences (including tariffs), the objectives (regulate and provide incentives for efficiency), the role of arbitrator between the administration and companies, and finally the basic role in defending the interests of users. This is all to be carried out transparently and independently. The regulator must not depend on political power or on companies. It must be working on behalf of users, so that they are provided with the best possible service at the lowest cost. And in order to achieve this, following the example of the Spanish energy regulatory system is advised (in addition to considering other experiences in the urban water industry). Although this deals with another essential resource (energy), it has the advantage of being within the same geographical context.

Chapter 1

The need for the regulation of water services key factors involved

Enrique Cabrera Jr.
Universitat Politècnica de València

1.1 THE NEED FOR REGULATION IN THE WATER SECTOR

From the moment of general appearance in the 19th century, urban drinking water services have been natural monopolies. This situation arises from the need to immobilise a large amount of capital in infrastructures and the intrinsic difficulty of installing more than one parallel network. For similar reasons, many public services (electricity, telephone, gas) were also natural monopolies for many decades. In other words, the service was provided by a single organisation (private or public) without users having any other alternative.

In recent decades many of the aforementioned monopolies have ended and, thanks to the opening up of these sectors and the proactive role by public administrations, these services are now provided in markets where there is competition[1]. Nevertheless, many water services are resisting this trend at this point well into the 21st century in a market where service providers in urban environments are still the exclusive operators.

The disadvantages for citizens in a monopolistic market are fairly obvious. On the one hand there is no natural incentive for the service provider to be competitive. The quality of the service does not need to be any better than a competitor's, and neither do prices need to be lower. Demand is captivated and therefore guaranteed. This statement is doubly true in the case of drinking water services as they are an

---

[1]In the case of telecommunications, competition in many cases is pure and users can choose between two or more different infrastructures providing them with the same service. On other occasions regulating the market has forced infrastructure owners to share their infrastructures with other companies in the same sector (for example this is the case for fibre optics, the electricity grid or the natural gas network in Spain).

indispensable element, not only for development and human welfare, but for life itself[2]. In economic terms, the demand for water services is inelastic, and therefore not tightly subject to traditional market forces. This is therefore an undesirable situation in which the service provider is in control of practically all the market variables.

Consequently, in the case of water services (as in many other public services) the administration needs to act in order to defend the interests of citizens and regulate the market in order to avoid possible situations of vulnerabilities of citizens to unethical practices stemming from the monopoly of these services.

Service regulation centres on three fundamental pillars (Malyshev, 2010):

(1) Control of entrance on the market.
(2) Price control.
(3) Quality control.

In other words, on the one hand regulators should determine which companies or organisations are allowed to provide the regulated services in specific areas.

On the other hand, they should ensure that fair tariffs for the service are established, guaranteeing long-term sustainability of the infrastructures (in the case of private companies), and a reasonable profit that is justified by efficient operation of the service.

Finally, it is fundamental to exercise quality control over the service users receive, since there is no incentive other than this control (and the willingness of the service provider to do a good job) which can guarantee adequate levels of service.

It is obvious that controlling prices and quality are closely linked (since a fair price will depend on the product quality, namely water quality, and the service offered). Nevertheless, the necessary knowledge and tools to exercise control of the two pillars can entail significant differences. Price control means economic regulation of the services, whereas quality control entails technical regulation.

It is rather odd how regulation in the water sector often arrives after regulation of other public services such as telecommunications or energy, bearing in mind that water has been recognised as a human right. More so when we observe how regulation of water services is often designed on the basis of experience acquired from regulating the energy sector. This practice, although understandable, could lead to unsuitable practices for a sector with its own specific idiosyncrasy.

This said, there are several factors that differentiate urban water services from other public services. Understanding and assuming these differences is essential when establishing the mechanisms that govern the sector, and establishing the necessary differences between regulating this service and other services.

---

[2]We should not overlook the fact that a lack of drinking water and sanitation is the cause of a great many deaths in the world each year.

1.2 SINGULARITIES OF URBAN WATER SERVICES

Urban water services are different from other public services and, what is more, those differences are critical when establishing the mechanisms and tools that permit regulating them. In other words, there are singularities regarding water services that advise designing specific mechanisms and tools for them.

These differences are listed as follows:

(a) *Water services cover a human right.* The first difference between any other public service and water and sanitation services is that the latter is a means to satisfy a human right recognised by the United Nations Assembly (United Nations, 2011). More specifically, in point 7 d), the resolution urges states to "*Assess if the current legislative framework and policies are in agreement with the right to drinking water and sanitation and to detract from, amend or adapt them, as relevant, to guarantee compliance with the principles and standards of human rights.*"

In other words, regulating drinking water and sanitation services must take that right into account, considering the principles of availability, quality, acceptability, accessibility and affordability.

Historically the nature of the fundamental right to water has been reflected in specific legislation for these services in many parts of the world (in many countries the drinking water service cannot be cut off despite payment defaults). It is therefore obvious that any attempts to regulate the sector should be based on this basic premise.

(b) *Water is local.* During live presentations, my transparency concerning this point is titled "water is heavy". Almost certainly because of its fluid property, we often forget the specific weight of water, making it denser than the material making up the human body (which is why we float in water).

When we value drinking water supply services, we hardly ever bring attention to the amount of mass that is displaced in order to meet the needs of users. And nevertheless, there are very few services (or none at all) that entail regular delivery of a quantity of material that even approaches the quantity of water a user consumes on a daily basis.

The estimated weight of the service provided to a Spanish family of 4 members[3] of different public services is shown in Table 1.1. As can be seen, consumption of 130 l/person/day (as per the National Statistics Institute [INE], 2013) gives a figure of around 200 tonnes of water delivered every year to each home of four members. In other words, as can be seen in the table, water is heavy, very heavy.

---

[3]Average consumption of 130 l/person/day has been used for calculations, and natural gas consumption of 2.5 kWh/day.

Table 1.1 Weight of annual supply of different public services to a family of 4 members.

Service	Weight of Annual Supply
Electricity	0 kg
Natural Gas	55 kg
Water	189,800 kg

This difference in the mass to be transported for each service is a determining factor when establishing the nature of each of them. Whereas electricity and gas transport is carried out across thousands of kilometres without any problem (international transport of gas is common from Algeria and Russia towards western Europe, and also electricity is transported from France to Spain for example), transport of water, although technically possible, is much more expensive in terms of energy whilst also requiring larger infrastructures.

Therefore water is usually supplied locally (and, with a few exceptions, is from sources located within a radius of just a few hundred kilometres, at the most, from the final point of consumption). Nevertheless, users often use gas or electricity that has originated several thousand kilometres away.

The importance of this fact is higher than is initially apparent. All told, by introducing competition in the retail supply (end users) of water, the situation becomes more complicated by not being able to establish a uniform price for the resource, not even at local level. Depending on the source of the supply (surface, underground, desalination, etc.) the cost of the resource can be significantly different. On the other hand, the operating costs themselves can also vary greatly depending on local conditions (in places with abundant water supplies reducing leaks is less critical; in places with more abrupt topology, pressure needs to be higher, resulting in greater losses). This is therefore a relevant factor when regulating water and sanitation services.

(c) *Water services are complex*. As we pointed out in the introduction, one of the fundamental pillars of regulation is controlling the quality of the service. In this sense it is important to emphasise the role that users' expectations play. The standards ISO 24500 (ISO, 2007a, b, c) pursuant to assessing drinking water and sanitation services, establish that assessment of these services must be performed in accordance with the needs and expectations of users.

Said expectations are the third major difference between public water services and public energy services, for example. Whereas in the electricity or gas services supply of the product is practically digital (except for one or two parameters, end users do not perceive the quality of the "product"), the case of water is very different.

Figure 1.1 shows how urban water services have a higher number of parameters that influence the perception of the service by users. On the one hand the quality of water as the "product" is much more variable and at the same time can be perceived in detail by users (smell, taste, colour, potability). On the other hand, the service also has an impact that directly affects users (flooding, environmental impact of waste water).

An anecdotal example, although significant at the same time, showing that supplying drinking water is complex and concerns users, are the auxiliary services and products that have appeared around water. Consumers often fit descalers or reverse osmosis membranes to improve potability of water, hardness or the organoleptic properties of the water they consume. There is technology available today that would allow service providers to improve these properties before supplying water, and therefore establish a different level of service. Rather curiously, the local nature of water also has an influence on users' perception of the service. It is very common, even in the same supply, to have very different water quality depending on the source of the water. Once again, electricity and gas services, often regulated parallel to water services by the same body, do not have this particularity.

Basically, urban water services entail a number of singular features that deserve to be taken into account during regulation. On the one hand, economic control of the services must take into account the fact that it is a human right and the cost differences that the local nature of water confers. On the other hand, when establishing suitable levels of service, it will be necessary to control and guarantee a higher number of variables. Whichever is the case, what is clear is that it is not enough to simply regulate water services without employing other tools than those directly imported from regulation of other sectors such as energy or telecommunications, and it will be necessary to establish mechanisms exclusively inherent to the water industry.

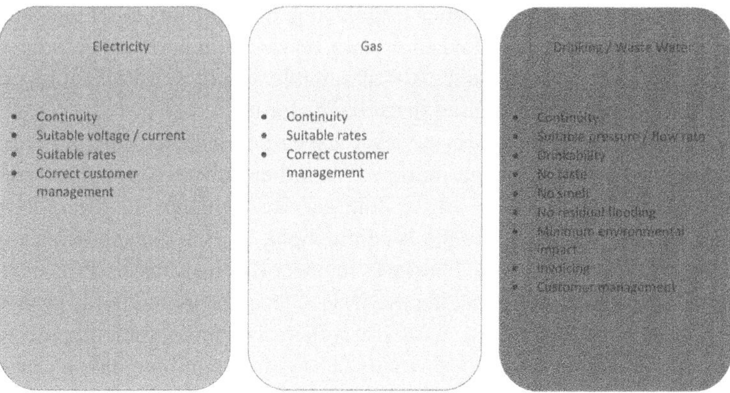

Figure 1.1 Comparison of user expectations for public electricity, gas and drinking/waste water services.

1.3 SERVICE LEVELS

Water and sanitation services can be provided at different standards of service. It is commonplace to find situations, in countries where these services are regulated, where drinking water supply is intermittent and a significant part of the population does not have access to drinking water and/or sanitation services. In other countries however, regulation centres on aspects such as the environmental impact of the service or the the time for which a call to the customer service centre is on hold.

These differences are not present by chance. Water services have evolved over time, and most of them pass through three stages defined by Lobato de Faria/ Alegre (1996).

- *Quantity stage:* Where the objective is to meet the biophysical needs.
- *Quality stage:* Where, in conjunction with the need to supply sufficient quantities of water, other psychological, cultural and aesthetic conditioning factors are added.
- *Excellence stage:* With the appearance of social, economic and environmental sustainability.

Depending on the historical, social and economic circumstances of each place, the service level offered by the provider can vary, and furthermore so may the service level required or desired by the administration.

It is fairly obvious that offering a high service level entails higher costs than a basic service. This principle, apparently so simple, often seems to be overlooked when establishing mechanisms to determine the price for urban water services. Hence it is possible to find simplified cost models that only consider variables such as the length of the network or number of delivery points, but which do not take the quality of the service provided into account, and therefore the calculations concerning those costs are unlikely to include all the conditioning factors.

It is therefore indispensable when establishing mechanisms to set fair tariffs for a service, that the tariff is linked to certain service standards, since otherwise it will be impossible to qualify said tariff as suitable or fair, as it will not be possible to determine what is being offered in exchange for it.

In a market economy, the service level a specific company must provide to its customers is established by the market forces themselves. Companies are often able to find niches providing an existing service although at different service levels and prices. In many sectors it is commonplace to see the appearance of new competitors under the "low cost" business model, offering a basic service at a much lower price in exchange for reducing the service level compared to the pre-existing offer by established companies. Likewise, other companies sometimes centre on a "Premium" business model in order to satisfy the more demanding consumers in exchange for charging a higher price than their competitors. In all these cases users decide which alternative they prefer and are free to choose the company that provides one service or another.

Owing to the natural monopoly of water and sanitation services, the service level cannot be established by market mechanisms and needs to be established by the administration department in charge of regulating it. In that context it would appear logical to constantly glean the consumers' opinions about their preferences concerning the service, as is the case for the Consumer Council for Water (CCW)[4] in the United Kingdom.

It is precisely in this context that the ISO 24510 standard *"Activities related to drinking water and waste water services – Directives for assessing and improving services for users"* (ISO, 2008a) acquires more importance. This is a non-certifiable standard, which means that operators are not awarded a compliance seal accrediting compliance, which is probably why it is not particularly popular. Standard 24510 pertains to a family of three international standards aimed at assessing and improving urban water services and was the result of the work by a committee of 100 experts involving more than 30 countries widely representing all the players involved in water services[5].

The specific standard ISO 24510 specifies that the management objectives of urban water services must be established in accordance with the expectations and needs of the users. And in spite of the fact that the standard does not define any mechanisms for regulating services (that is not its function), it would be logical to apply the same principles when establishing the objectives of regulation.

The document includes a compilation of the requirements users demand for their drinking water and waste water services. It is a thorough list, but one that remains open, and can be used as a baseline to establish user priorities for a specific service.

More specifically, the needs and expectations of users listed in ISO 24510 are as follows:

- *Access to water services*
- *Service providing*
  - New connection time
  - Repairs
  - Reasonable service price
  - Quantity of drinking water supplied
  - Drinking water quality
  - Aesthetic aspects of water

---

[4]The CCW is one of the regulatory structural pillars in England and Wales whose function is: *"to represent consumers of water and waste water services (… and) to ensure the collective voice of users is heard in the national debate on water and that consumers are kept at the heart of the water industry"*. As can be seen on their website (http://www.ccwater.org.uk) the CCW has offices throughout England and Wales and deals with complaints from users that have not be resolved by the water companies.

[5]Despite the fact that the experts on ISO technical committees are part of the delegations from participating countries, their contributions are made as individuals. Public and private operators, academics, consultants, representatives from administrations and consumers all contributed to the drafting of standards for ISO TC224.

- Pressure
- Continuity of supply
- Cover of services
- Flooding from waste water
- *Contract management and invoicing*
  - Availability of a clear service agreement
  - Invoicing accuracy
  - Response to complaints
  - Invoicing clarity
  - Payment methods
- *Relations with users*
  - Contact/complaints in writing, by telephone, visits
  - Warnings about restrictions and stoppages
  - Community participation
  - User participation
- *Environmental protection*
  - Sustainable use of natural resources
  - Waste water treatment
  - Environmental impact
- *Safety and emergency management*

Each item on the preceding list must be considered in each service, for which a target or minimum level must be established and which must be reached. In fact, the standard does not only describe users' expectations and needs, but also provides possible assessment criteria and methodology for each of them which should be followed in order to build or select indicators that permit monitoring compliance, as shown in Figure 1.2.

It is important to point out that ISO 24510 does not contain any target service levels or numerical references. On the one hand, the technical committee in charge of it did not consider that this was the role of the standard. On the other, it is very complicated to establish single reference values for the entire world[6]. That is a task that regulators should perform, but one they can do following the guidelines established by the aforementioned international standard.

It is important to emphasise that achieving certain levels of service entails costs. Therefore, when establishing a standard for urban water services, the tariffs users are to pay for said services[7] to recover the resulting costs are in some way being

---

[6]Difficult perhaps, but not impossible. In fact in AquaRating (Krause *et al.* 2015), the recent international system for rating drinking or sanitation water services does establish numerical targets that operators should maintain if they want to obtain the maximum possible score.

[7]Under equal conditions of efficiency, investments, personnel and operating costs will be higher or lower depending on the quality of the service to be offered. Reducing lead times, improving water quality, reducing risks of infrastructure failure, etc., all entail cost overruns which ultimately need to be recovered through service tariffs.

indirectly conditioned. In other words, it is practically impossible to assess services from the economic point of view without taking into account the quality level at which the service is being provided, although the opposite is not necessarily the case.

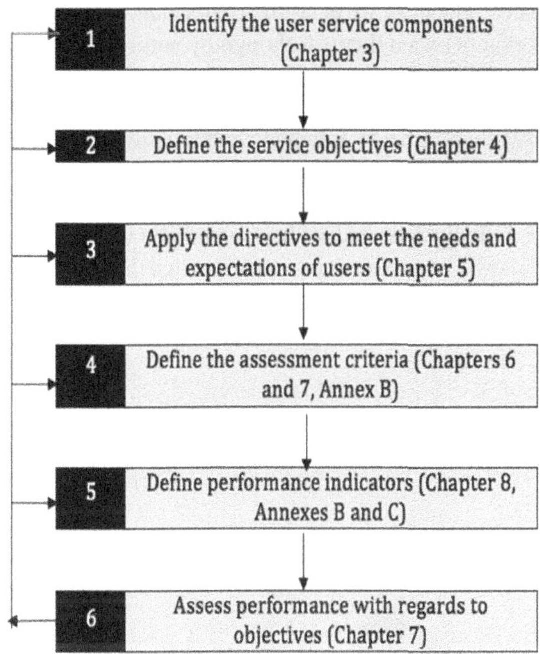

Figure 1.2 Content and application of standard ISO 24510 (ISO, 2008a).

1.4 TECHNICAL REGULATION AND ECONOMIC REGULATION

It is common to find examples that differentiate between so-called technical regulation (or service quality) and economic regulation in literature and in reality. In line with the regulation pillars established previously, price control of the services comprises the so-called economic regulation, whereas quality control of the service is the technical regulation. Many of the regulatory bodies of water and sanitation services in the world do not exercise the functions of economic regulation (Marques, 2011) but only exercise regulation of service quality.

The best example to date is probably the Portuguese regulator, ERSAR, created in 1997 (originally IRAR) to regulate drinking water, sanitation and solid waste services. During the first years of its existence, ERSAR exclusively regulated service quality, in the way that has become known as sunshine regulation. This type of regulation does not consider economic incentives for companies that meet the service standards demanded by regulators. In view of the absence of sanctions or economic consequences, these types of regulators need to find other items for motivation, which are based on the use of public communication and tools of the so-called *yardstick competition*[8].

Most independent water services regulators follow the path laid out by OFWAT, using performance indicators to measure performance by operators and create artificial competition between them. Although by nature a regulator is transparent, in "sunshine regulation" disclosure of the results achieved by operators becomes the main tool to motivate service providers to improve, known as *naming and shaming*.

Rather curiously, before being assigned economic regulation attributes, ERSAR achieved excellent results with this approach. This approach entails teaching citizens about urban water services and arousing interest among users about the level of services they are actually provided with. An example of this approach is given in Figure 1.3 which shows the application for mobiles that the regulator has created, thus meaning anybody can consult the quality of the services controlled by the regulator.

Figure 1.3 Mobile application created by ERSAR to inform users about Portugal's water, sanitation and solid waste service levels.

---

[8]Regulators compile indicator values from different operators and compare them to each other thus creating a competitive environment which does not really exist because each operator's market is independent from the rest.

Technical regulation does not exclude de facto existence of some kind of economic regulation to guarantee fair prices, in line with the service levels. This function often lies with the municipal authorities, who in many cases have to approve the water service tariffs. Nevertheless, one of the main flaws in places where quality control of services and tariff control are not in the same hands, is that tariff calculations do not cover all necessary variables. Consequently, the drawback of having a regulator who is only in charge of controlling the quality of the service is not the regulator itself, but rather the administration that establishes the service tariffs, and which, to do so, should take into account all the necessary technical criteria. Unfortunately, local authorities do not normally take all these factors into account.

Whichever is the case, what is obvious, and common practice by nearly all regulators, is that it is necessary to assess and control performance by operators in order to be able to regulate service quality. Performance indicators, and more specifically those developed in accordance with the criteria established by the International Water Association (IWA) are the ideal tool to carry out this function (Alegre *et al.* 2016). The aforementioned criteria are largely shared by the ISO 24500 standards mentioned previously. The IWA and ISO 24500 standards establish (see Figure 1.2) that the indicators used to assess performance must meet previously established objectives. When used for regulation, those objectives must be the same as the regulator's.

Marques (2011) lists the objectives for regulation of urban water services:

- To protect the interests of users in terms of obligations of a public service
- To promote efficiency and innovation
- To ensure stability, sustainability and solidity of water and sanitation services

We shall now move on to discuss each of the aforementioned objectives in more detail.

1.4.1 Protecting the interests of users in terms of obligations of a public service

As mentioned in the introduction to this chapter, there are a number of goals that public water services should achieve which have even been included in different rights, objectives and documents by the United Nations. These include universality of the service, equality, accessibility and protection of health.

Additionally, continuity and service quality, assurance of assets and services, transparency, option of different payment methods, representativeness and participation by users in decision making and the existence of conciliation and resolution of conflict mechanisms should also be included.

All these objectives are part of the list of expectations and needs of users listed in ISO 24510.

Some of the conflicts between these commitments and the need by some private operators to maximise the return on investment by their shareholders are obvious;

and the role of the regulators as referees to allow the different players to achieve their goals is absolutely critical.

A clear example of the need for a regulator is in providing services to remote areas where the cost of increasing the cover of the service by very small percentages is considerable and is not sustainable from a purely business perspective.

1.4.2 Promoting efficiency and innovation

One of the problems stemming from monopolies is the lack of incentives for innovation. Whereas in a free market innovation is part of the strategy for possible success, in a natural monopoly the only reason to innovate could be to improve internal efficiency but not to improve the service or the product. In the case of publicly managed operators, the incentive to be more efficient could be lower still, since there is no need to make a business profit.

In the words of the first Scottish regulator:

"Ultimately the best way to promote the interests of consumers in a public sector model is by improving the economic efficiency of the industry, and therefore the value of the money produced" (Water Industry commissioner for Scotland [WICS], 2002).

Whatever the case, the very concept of efficiency relates the inputs and outputs of a system. In the case of water services, in order to determine efficiency the quality of the provided service needs to be taken into account to determine whether certain inputs are necessary for such service levels.

1.4.3 Ensuring stability, sustainability and solidity of water and sanitation services

One of the major problems in urban water services is their long-term sustainability, mainly the infrastructures. This problem is widely generalised in Spain, but also fairly widespread in the rest of the world. Most of the water pipes in our cities were installed many decades ago (some over 100 years ago) and the average age increases year after year (whereas their conditions deteriorate accordingly). Figure 1.4 shows the average age of drinking water infrastructure in the United States related to time, reporting a clearly unsustainable progression.

Since the distribution networks (and the sanitation and drainage networks) are the critical component for providing services, but also the most difficult and costly to replace (not even with infinite resources would it be possible to replace the entire distribution network in the short term), this evolution suggests that the continuity of the service as we know it today is in danger if the situation is allowed to continue. Experience has taught us that networks that are not properly maintained start to leak excessively, having a major impact on the resources and the quality of the water. During times of water shortage this can lead to interruptions in the supply and even faster deterioration of the condition of the network due to regular pressure increases and decreases in the system.

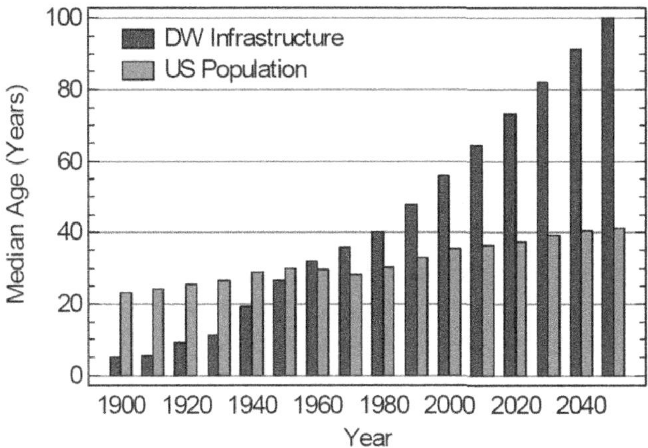

Figure 1.4 Evolution of the average age of drinking water infrastructure – blue series – compared to the average age of the population in the United States (Buchberger, 2011).

The fact that this is a very widespread problem has not occurred by chance. On the one hand existing water tariffs in most of the world do not account for the costs of replacing the infrastructure. Therefore, income from invoicing for the service only serves to cover the operating costs, but not the tied-up capital.

In view of this situation, private companies (who are not the owners of the infrastructures except for a few exceptions, the most notable being England and Wales) have limited incentives to maintain the networks in perfect condition. The tariffs do not permit this, and applications have to be made to the owner of the infrastructures (the town) for extraordinary action to permit this work (in the form of subsidies or tariff increases). This means that renewing the network becomes the exception on many occasions, when in all reality it should be ongoing, scheduled action.

On the other hand, the lack of extraordinary grants, in cases of short duration concession contracts and when there is uncertainty about continuity, investments in the network only reduce the profit margin. Moreover, it is not uncommon to hear certain operators say that they are not going to improve the network at the end of the contract so that the benefits of a better infrastructure are exploited by the next concessionaire.

Finally, part of the responsibility for ageing infrastructures has to be attributed to the owners themselves (the towns). Investment in buried infrastructures is a decision that is difficult to justify in short-term politics. Renewing pipes requires a high investment. This is an investment that is not visible and where the immediate effects are not noticed by anyone. Furthermore, in the case of future problems, more often than not the people responsible are not even there to suffer from or deal with them, in addition to the fact that they are able to dilute the responsibility of renewing the pipes among all of those that preceded them during their lifespan.

That is why the figure of a regulatory body is indispensable, regardless of local politics, as long-term plans for water services are necessary. Hence, in an ideal world the average age of networks should not permanently undergo excessive variations. In other words, networks should be capable of operating equally well (or better) within several decades.

This concept of sustainability of the infrastructure is now present in many forums that deal with urban water. The widespread problem has led to numerous projects related to Heritage Management of Infrastructures, as can be seen in the European projects AWARE-P (www.aware-p.org) and TRUST (www.trust-i.net). Both projects provide free tools and materials developed in order to help achieve sustainability of urban water services.

A final aspect regulators must ensure is the financial sustainability of the service. In the current scenario of economic crisis it is hardly surprising that regulators have to guarantee that the level of debt incurred by companies providing the services is within reasonable margins. This type of sustainability becomes even more important in countries where infrastructures belong to the supply companies (England and Wales), where bankruptcy of the companies would result in serious problems with regard to guaranteeing the service.

In the case of Spain this problem does not arise with the private companies, but could be relevant for public companies or municipal services. In this sense, the fundamental task of the regulator should focus on ensuring the revenue from the service is sufficient to guarantee economic sustainability (and in the case of private companies, it should guarantee a reasonable profit margin).

A clear example of the foregoing arises when consumption of a supply is significantly reduced (for example in efficient water use programmes). Paradoxically to what could be initially thought, this reduction in consumption by users should be accompanied by an increase in tariffs in order to maintain the operator's income (on the basis that the previous tariffs were actually reasonable).

Basically, the regulator is the player in charge of many conditioning factors involved in modern provision of water and sanitation services, being brought together in a sustainable solution that is adequate for all parties, but particularly for the users of the service. The regulator is an independent player from all the rest who should coordinate interaction between users, service suppliers, local authorities and the administration departments at higher levels.

1.5 BASICS OF TECHNICAL REGULATION OF WATER SERVICES

The available methodologies to carry out economic regulation of the water industry have been described in depth in literature. The similarities between these methods and those used to regulate other sectors (energy, telecommunications, etc.) has led to a wide choice of books and research articles that detail the use of econometric tools to establish tariffs in a regulatory framework for water services.

Nevertheless, as described in section 1.2, urban water services entail unavoidable singularities that must be addressed when defining regulations. The higher gradation in terms of possible service levels (compared to other sectors there are many more possible levels of service in urban water services) demands that special interest is paid to assessment and consideration when contemplating the economic side of regulation. This section attempts to contribute to complementing the current scenario in terms of tools for regulation, which in general is still dominated by those used for economic regulation.

Establishing objectives for services levels provided to users and mechanisms to control them is the essence of technical regulation for water services. Simplifying concepts, the aim is to guarantee that the service is provided adequately and in accordance with required standards.

Complying with the task of establishing objectives and control requires a tool that permits measuring services levels. Said tool consists of performance indicators[9].

Performance indicators in the urban water sector have been in systematic use since the nineties. In fact it was OFWAT who brought attention to the concept which had been in use in the regulatory system in England and Wales right from the start. Nevertheless, publication of the Best Practices Manual by the IWA (Alegre *et al.* 2000) gave a major boost to the use of indicators in the industry, with the third edition scheduled to be published in 2016.

The IWA system has become the main benchmark in the urban water industry for establishing performance indicator systems, regardless of the purpose of them. The ISO 24,500 standards and international systems such as IBNET (International Benchmarking Network, Danilenko *et al.* 2014) or AquaRating (Krause *et al.* 2015) also took the directives established by the IWA as their baseline.

The IWA indicators system clearly defines three basic parts of a performance indicator system:

(1) On the one hand, the IWA system (particularly as of the second edition of 2006) accurately defines the structure, the components and the vocabulary associated with a system of performance indicators. Hence, the hierarchy of an indicators system ranges from the *variables* (which comprise the basic source data used to calculate the indicator values), to the performance indicators themselves or *context information* (which serves to explain differences in the services offered due to exogenous factors not stemming from service management itself – such as the network topology, which has an influence on pressure and hence water losses).

---

[9]Regardless of the methods used to interpret them, most water service regulators in the world use performance indicator systems to control operation of services.

(2) On the other hand, the IWA system provides a list of over 100 indicators and more than 200 variables. Each of these items has been defined in detail and has been finely tuned (when necessary) in each subsequent edition of the indicators system. Thanks to this process, they have become standard measurements for certain assessment criteria. For example, the water loss indicators defined by the IWA are considered as optimum to assess this area of performance. That does not mean that the IWA indicators are valid in any situation or context, but it is advisable to study the options afforded by the IWA for a specific indicator before defining a measurement from zero.

(3) Finally, the IWA manual confers due importance to the management of data quality in a system of performance indicators and provides the methodology to put it into practice. In general terms, when the value of an indicator is shown in comparison to the results of other operators (reminding us of "yardstick competition"), it is common practice to dispense with any qualifier regarding the source of the data. Nevertheless, the quality of the data in the water industry is notoriously improvable. The bottom line is that most of the infrastructures are inaccessible, underground and have an average age spanning decades, sometimes even a century old.

Faced with identical values of real losses in two different systems, minimal critical analysis would reveal that it is impossible for both numbers to have been obtained with equal rigour or certainty. The deviation of the value shown from the real value (which is always unknown) will largely depend on the number of meters and flow meters, simultaneous readings, the accuracy of the measuring equipment (conditioned by quality, technology and age) and the estimate of unmeasured flows, if any. Nevertheless, in a system of performance indicators that does not take data quality into account, both values will be shown next to each other suggesting identical performance in both systems, which is almost certainly not true[10].

This aspect, critical in any performance assessment system, is even more critical in the case of systems on which regulation is based. This was recognised by OFWAT in its day, establishing a data quality record system which served as the basis for publishing the indicators in the first IWA manual.

Over time the OFWAT system (based on the concepts of reliability and accuracy) has proved to be too complex for a large number of real applications, particularly when assessment of data quality is carried out by the operators themselves. Therefore, the proposal by the IWA has included simplified options to record the quality of the data (with a view to at least these options being used in all indicator systems). In the third version of the IWA system (Alegre *et al.* 2016) there is still a version of the OFWAT system as the ideal alternative, but more simple alternatives are also afforded.

---

[10]An extreme case could arise when one of the systems has all its volumes recorded correctly and the other resorts to estimating most or all of them. In this scenario how could the same value for an indicator account for similar performance?

In recent times, indicator systems have proved to be insufficient to assess certain concepts. On the one hand, there are aspects of the service that cannot really be compared or even measured. Among the difficulties involved in establishing indicators that permit assessing certain aspects, are the following:

(a) There is not enough data or the quality of the data is very poor. Although it would be desirable for performance assessment with indicators to feature better data quality than the data currently obtained from the systems, the truth is that in practice it is common for some indicators not to be particularly useful in certain contexts because it is impossible to obtain quality data. A clear example of this (following on from the example of losses) would be a context in which none of the systems is fitted with water meters at delivery points, and therefore all the delivered volumes would be estimates. Another common circumstance in some countries is not knowing the actual length of the network or the number of delivery points fitted to it. In these cases establishing an indicator that depends on these variables (particularly when comparing values for regulatory purposes) makes no sense.

(b) The concept to be "measured" cannot be measured or assessed throughout the network with indicators. This is the case when assessing desirable values such as transparency or governance. Although it is possible to perform a theoretical exercise to attempt to build indicators that measure some aspects of these concepts, the truth of the matter is that the reality reflected by those indicators will always be somewhat limited.

(c) There is an indicator to measure the concept but the context prevents a balanced or useful comparison. A common example of this is the impact the service has on the resource, for example the abstraction of water compared to the available resources. Even though it is possible to find indicators that assess the concept, comparison will not give a clear view of the reality concerning the concept (for example, in the case of systems that are supplied from surface, underground or even desalinated water, it would be very difficult to make comparisons of different systems with combinations of these resources and establish the degree of sustainability of each of them, which is more harmful for the environment depending on the use of the resource).

(d) There is an indicator available but the information feeding said indicator cannot be verified, or at least not at a reasonable cost. An indicator could be perfect, but it may not be possible to verify its value in an independent way, or verification would entail excessively high costs with regard to the available resources.

(e) There is an indicator to measure the item but comparing it between two different systems is impossible because the context (physical, social, political, economic) notably conditions the values reported for the indicator.

These limitations are particularly critical when they appear in systems that should help to regulate water services. Including indicators to assess items of this kind will only give a partial or incorrect view of reality. Nevertheless, the absence of any kind of measurement will mean that the regulator does not measure the operators' performance in that field, and therefore will not apply the advantages of benchmarking and artificial competition between operators.

An alternative in the absence of valid indicators is assessing practices, a method that has been developed considerably in the AqauRating system (Krause *et al.* 2015). The concept of assessment of practices is relatively simple and effective. This system tries to focus assessment on processes, best practices and achievement of certain milestones through questions that can only be answered yes or no, without defining any measurements of the results.

A clear example illustrating the usefulness of practices is the complicated subject of assessing how access to the service is being guaranteed. It is a problem that does not only depend on the operator, but is often related to municipal politics. Table 1.2 shows six practices reviewed under the AquaRating system in order to assess this item.

The disadvantage of this assessment by practices is that it does not guarantee service performance reaches a certain level. Nevertheless, the use of practices also has different advantages:

- For operators it means they are able to give reliable answers in the process much quicker and more conclusively. Obtaining a reliable value for a variable (e.g. the population subject to drinking water services) is often a complicated task shedding results that always involve a certain degree of uncertainty. Deciding whether to answer using "yes" or "no" is always easier.
- Likewise, it is much easier to categorically state if a certain level of reliability is achieved in the data. For example, it is extremely difficult to guarantee uncertainty in measurements of water meters (e.g. guaranteeing that the margin for error is below 10%). Nevertheless, it is easier to check if monitoring of stopped meters is carried out, whether or not a policy is established to renew them and whether there are procedures to estimate the error in the meters.
- Practices permit making progress in future performance. An operator can be making major efforts (e.g. renewing the network) that are not reflected in the value of some indicators until some time in the future. Nevertheless, the fact that a number of best practices are being implemented can be useful for assessment.
- Sometimes practices can identify if a minimum level has been reached. If regulators want to reach that level, but do not give incentives for improvements beyond that (it may not be efficient), it is not necessary to measure using indicators, but simply to establish a yes/no question about whether the minimum levels have been reached.

Table 1.2 Item AS1.1. of AquaRating – guaranteeing "access" to the service. (Krause *et al.*, 2015)

1	There are plans to extend the drinking water service to homes that currently do not have "access" at home, with established coverage goals and deadlines, which are equal to or even more demanding than those established by the relevant authority. (The practice is considered accomplished with maximum reliability if coverage of inhabitants with home connection to the drinking water network in the "territorial scope to be rated" is higher than 99.9% in the previous year to rating, if this is justified by data published by a "competent official body" or by the value of AS1.2 (>99.9%) supported by average reliability of 0.8 or more).
2	There are drinking water supply programmes through alternatives means to areas without home service in the "territorial scope to be rated" by the service provider for drinking water, guaranteeing availability at distances lower than 500 m for each home, continuity and quality of the service. (The practice is considered accomplished with maximum reliability if coverage of inhabitants with home connection to the drinking water network in the "territorial scope to be rated" is higher than 99.9% in the previous year to rating, if this is justified by data published by a "competent official body" or by the value of AS1.2 (>99.9%) supported by average reliability of 0.8 or more).
3	There are plans to extend the drainage service to homes currently with no connection to the mains "system", with established coverage goals and deadlines, which are equal to or more demanding than those established by the relevant authority. (The practice is considered accomplished with maximum reliability if coverage of inhabitants with home connection to the mains drainage "system" in the "territorial scope to be rated" is higher than 99.9% in the previous year to rating, if this is justified by data published by a "competent official body" or by the value of AS1.3 (>99.9%) supported by average reliability of 0.8 or more).
4	There is a special tariff system or a subsidy programme for the low-income population, to facilitate payment for normal consumption of drinking water and/or sanitation services.
5	There is a special tariff system or a subsidy programme for the low-income population, to facilitate payment for normal consumption of drinking water and/or sanitation services, either through subsidisation of this payment or through credit. (The practice is considered accomplished with maximum reliability if coverage of inhabitants with home connection to the drinking water "system" and with home connection to the drainage "system" in the "territorial scope to be rated" is higher than 99.9% in the previous year to rating, if this is justified by data published by a "competent official body" or by the value of AS1.2 and AS1.3 (>99.9%) supported by average reliability of 0.8 or more).
6	There is a department or specific operating area for planning and dealing with areas without drinking water and/or sanitation services. (The practice is considered accomplished with maximum reliability if coverage of inhabitants with home connection to the drinking water "system" and with home connection to the drainage "system" in the "territorial scope to be rated" is higher than 99.9% in previous year to rating, if this is justified by data published by a "competent official body" or by the value of AS1.2 and AS1.3 (>99.9%) supported by average reliability of 0.8 or more).

Assessment through practices is an effective alternative that is complementary to the traditional use of indicators. Despite the fact that the use of practices is not widespread among water service regulators, they are very likely to be used in the future to complement current performance assessment systems.

1.6 ANALYSIS OF PERFORMANCE ASSESSMENT

Even the best performance assessment system can be completely useless if the values of the indicators and other assessment items are not analysed properly and correct conclusions are not drawn.

Unfortunately, performance assessment is not an exact science and less so when behaviour of different services is compared. Specific conditions can have a major influence over the ability to reach certain performance levels.

For example, it is not possible to demand the same energy efficiency from an operator managing a flat network, who receives gravity-fed water with hardly any pumping requirements, as from another operator drawing water from an underground source and providing services over rugged terrain. It is obvious that the energy costs in these two cases will be very different and a direct comparison between the two without taking the context into account will be worthless.

The need to take information into account about context when analysing performance indicator information has been well known for decades. Nevertheless, what has not been possible is establishing mechanisms that guarantee that the analysis of calculated indicators is unique and consistent regardless of the methodology employed or the person conducting the analysis.

This is undoubtedly one of the main weaknesses of indicators as a decision-making tool and for benchmarking. Nevertheless, there is no alternative approach to overcome this difficulty, and performance assessment systems based on indicators are still the most suitable for this task. But bearing in mind the task regulators have to carry out, this is undoubtedly a factor that needs to be taken into account.

All said, the role of a regulator who benchmarks performance is to be an impartial judge, which is rather unlikely when performance by operators working in very different contexts is compared. There are two fundamental options in the approach to performance assessment based on the values of a group of indicators. These two options are practically independent from the indicator system chosen previously:

(1) *Econometric models.* These are statistical models that have the objective of relating cause and effect. This way it is possible to establish whether or not a certain performance aspect arises out of the context of the service or out of management. OFWAT was also a pioneer in the use of these techniques, as described in a report by the Organisation for Economic Co-operation and Development (OECD, 2011): *"That is why OFWAT has developed sophisticated econometric models to benchmark the efficiency of regulated*

companies, whilst at the same time controlling the exogenous factors that can affect performance."

These models have the advantage of conducting a statistical analysis that does not depend on the opinions or knowledge of the personnel of the regulatory body. The models are based on what the performance should be according to the resources that are used to achieve it, and therefore are capable of establishing the efficiency of operators.

However, these methods are extremely complicated and only a few experts (mostly academics) can really understand how they work. Additionally, the models are only for the data actually used, and therefore the choice of indicators or parameters, and the quality of the data used, can notably affect the results that are obtained.

Perhaps that is why OFWAT (which recently changed the models it used to assess operators in England and Wales) has now decided to use a combination of models and calculate an average from the results. This decision leads us at least to question the results of those models. On the one hand, using several models implies that the results obtained will be different. Moreover, it has been stated that it is not known which model is better than the rest to assess efficiency. The solution to take the average does not seem particularly suitable either, and for many people has not been satisfactory[11].

(2) *Analytical assessment.* The second alternative is based on the human factor, and consists of assessing the values of indicators, supported by clustering techniques, filtering and graphical representation of the assessment items. At first sight it is a much less sophisticated technique, and can lead to bias by the people in charge of the assessment, whose experience, knowledge and judgement are critical factors to obtain an adequate assessment. Nevertheless, there are several factors that do actually make it recommendable:

- ○ *Transparency.* Unlike the statistical methods mentioned in the preceding point, analytical assessment of indicators has absolute traceability. All the conclusions can be justified and followed up on by a third party. There are no complex mathematical models or "black boxes" producing results from unknown sources.
- ○ *Simplicity.* The transparency of the method would not be particularly relevant if it were not relatively simple to understand the processes leading to a conclusion. In the case of econometric models, even when the details of the formulation and parameterisation of the models are known, analysing the virtues and flaws of an assessment is an extremely

---

[11] From the allegations the companies have submitted to Ofwat, often prepared by university experts hired for this purpose, it seems that the econometric and statistical methods lead to discrepancies in use and configuration.

complex affair. In analytical assessment of indicators there are no grey areas in the analysis.

- ○ *Flexibility*. The reality of water services is very complex. Statistical methods require a certain degree of parameterisation, whether direct (the factors affecting efficiency are defined) or indirect (the factors are not defined in the model, but in fact come from a limited population by previously choosing a group of indicators). Once the model is established it is then possible to modify it, but not without affecting all the assessments made until that moment. Analytical assessment permits considering additional factors and particularities of specific cases without affecting the rest of the assessment.

- ○ *Capability of considering data quality*. As mentioned previously, one of the frequent problems in the water industry is the absence of quality data. However, there are no references in literature about how statistical methods consider data with different levels of uncertainty and/or reliability. It is no less true that taking said information into account in an analytical assessment is also a complex affair[12]. Nonetheless, given the flexibility mentioned in the preceding point, this problem is at least addressable for analytical assessment and the person in charge can take the information regarding the quality of the data into account.

It cannot be denied that leaving one person or a group of people in charge of carrying out performance assessment on the basis of an indicators system can lead to doubts about the bias of the assessment. Nevertheless, this bias can also be included in the econometric models (by choosing the model, parameterisation, etc.) with the added disadvantage that the process used to reach the conclusions will be much less transparent.

To date many regulators, such as OFWAT, have trusted statistical models to establish measurements of efficiency, and there are numerous references in literature supporting this practice. However, regulators have also started to come under control[13] and the use of less sophisticated instruments for assessment could start to be considered an advantage over the use of other, less transparent models. Hence, regulators such as ERSAR chose a simple model in the past in which the regulator determines the suitability of performance without the need to resort to statistical models. By using this method, a significant part of the information regulators use to reach these conclusions is accessible for either of the parties who wish to make a parallel assessment.

---

[12]A dilemma would then arise, for example: Who performs better - an operator with good indicators but bad data, or an operator with good data and poor indicators?

[13]A British parliament commission recently determined that Ofwat had allowed its water companies to overcharge tariffs more than necessary, leading to a cost of 1200 million pounds for users (The Guardian, 2016).

1.7 CONCLUSIONS

Drinking water supply and sanitation services are a key factor in the lives of human beings and development of communities. Establishing these services responds to a human right, and something that is therefore essential for society at large. Owing to the natural monopoly inherent to these services, regulation needs to be established in order to guarantee that access to the management and service provision is transparent and fair, and that the quality of the provided service is suitable and the prices charged are in line with the quality of the service.

Nevertheless, and despite its importance, centralised regulation of urban water services is relatively new and in many places has not even been established in a uniform, structured way. It is common to find that regulations for water services are closely based on the regulatory models established for the energy sector. This model can and must be taken into account when developing a regulatory framework for water, but without forgetting that there are clear differentiating factors between both sectors. More specifically, controlling service levels acquires special importance owing to the number of variables involved in urban water services.

Despite this, in many parts of the world regulation of water services tends to forget about assessment of the service level as an integrated part of regulation. But if a regulatory framework is to be established to determine a suitable tariff structure, it is not possible to talk about costs without having first established what the services levels currently being provided actually are.

Performance indicators are a powerful tool for regulators to be able to assess compliance with service levels, which is known as technical regulation.

The results of using systems based on indicators depend to a large extent on the analysis that is made of the performance assessment. Traditionally, many regulators in the sector have chosen econometric models that attempt to establish if performance is efficient. These models may not be the most appropriate ones for the water sector bearing in mind the diversity in quality of the data and transparency when interpreting the results if we are not experts in these techniques.

Some regulators have started to use analytical performance assessment, which although dependant on the human factor, is completely transparent and traceable, and can therefore facilitate dialogue between the different players involved in the service.

1.8 REFERENCES

Alegre H., Hirner W., Baptista J.M. and Parena R. (2000). Performance Indicators for Water Supply Services. Manual of Best Practice Series. IWA Publishing, London, ISBN 1 900222 27 2, 160 pp.

Alegre H., Baptista J. M., Cabrera Jr., E., Cubillo F., Duarte P., Hirner W., Merkel W. and Parena, R. (2016, final draft). Performance Indicators for Water Supply Services, 3rd edn. Manual of Best Practice Series, IWA Publishing, London.

Buchberger S. (2011). Research & Education at the Nexus of Energy & Water. PowerPoint presentation.

Danilenko A., van den Berg C., Macheve B. and Moffit L. J. (2014). The IBNET Water Supply and Sanitation Blue Book 2014. The International Benchmarking Network for Water and Sanitation Utilities Databook. International Bank for Reconstruction and Development/The World Bank. ISBN 9781464802775

Instituto Nacional de Estadística, INE (2013). Estadística sobre el suministro y saneamiento del agua. http://www.ine.es/dyngs/INEbase/es/operacion

ISO – International Organization for Standardization. (2007a). ISO 24510: 2007. Activities relating to drinking water and wastewater services – Guidelines for the assessment and for the improvement of the service to users.

ISO – International Organization for Standardization. (2007b). ISO 24511: 2007. Activities relating to drinking water and wastewater services – Guidelines for the management of wastewater utilities and for the assessment of drinking water services.

ISO – International Organization for Standardization. (2007c). ISO 24512: 2007. Service activities relating to drinking water and wastewater – Guidelines for the management of drinking water utilities and for the assessment of drinking water services.

Krause M., Cabrera Jr., E., Cubillo F., Díaz C. and Ducci J. (2015). AquaRating: An International Standard for Assessing Water and Wastewater Services. IWA Publishing. ISBN 9781780407395

Malyshev N. (2010). Regulatory Frameworks in OECD countries and their Relevance for India. Presentación. http://www.oecd.org/gov/regulatory-policy/44933928.pdf (accessed November 2015).

Marques R. C. (2011). A Regulaçao dos serviços de abastecimiento de agua e de saneamiento de águas residuais. Uma perspectiva internacional. Instituto Regulador de Águas e Resíduos.

Naciones Unidas (2011). A/HRC/RES/18/1 – Resolución aprobada por el Consejo de Derechos Humanos 18/1. El derecho humano al agua potable y el saneamiento

OECD (2011). Meeting the challenges of financing water and sanitation. ISBN 9789264120518

The Guardian (Jan 13th, 2016). UK households overpaying for water supply due to regulator miscalculation. http://www.theguardian.com/money/2016/jan/13/ukhouseholds-overpaying-water-regulator-miscalculation

WICS (2002). Commissioner's Corporate Plan. Water Industry Commissioner for Scotland, UK.

Chapter 2

Portuguese regulatory model for water and waste services: an integrated approach

Jaime Melo Baptista

Chairman of the Board of the Portuguese Water and Waste Services Regulation Authority (ERSAR) Rua Tomás da Fonseca, Torre G - 8°, 1600-209 Lisbon, Portugal

2.1 HOW IMPORTANT ARE WATER AND WASTE SERVICES?

Drinking water supply, waste water management and solid waste management have taken on increasing importance in the global context, and are essential public services for the social and economic development of any country.

They have major implications for the environment and public health, and it can be verified that the healthiest societies with the longest life expectancy are those countries and regions with a high level of care regarding these services. On the other hand, the highest death and mortality rates occur mainly in countries and regions with an insufficient level of care regarding these services.

Hence, these sectors have been clearly earmarked as a priority by policy makers around the world and this had led to public opinion being increasingly attentive in this regard.

2.2 WHAT IS THE INTERNATIONAL FRAMEWORK?

Internationally there have been many policy initiatives to encourage governments to give greater attention to water and waste services.

Of note are the Millennium Development Goals adopted by the United Nations in 2000, which set targets for water services in terms of population coverage. This document set the goal of countries halving their population without access to safe drinking water and sanitation by 2015, which has not yet occurred in many situations.

Additionally, in 2010 the General Assembly of the United Nations declared access to safe drinking water and sanitation as essential human rights for the full enjoyment of life and for all other human rights, thus reinforcing the importance and concern that increasingly fall on these sectors.

These mean a right for all citizens to have access to water services which are appropriate and safe, essential to public health and the protection of the environment, through traditional collective systems, simplified collective systems or individual systems. The services, as human rights, must be physically accessible, appropriately sized, hygienically safe, affordable and culturally acceptable. Citizen participation in decisions, access to services without discrimination, monitoring of the situation and its regular reporting should be ensured.

These rights also mean that member countries of the United Nations have an obligation to carry out the necessary measures to achieve them. The pursuit of these rights by member states, through their governments, means the obligation to respect, not threatening nor limiting access, the obligation to protect, that is, prevent threats or limitations by third parties, including providers of water services, and the obligation to fulfil these rights by supporting citizens in their access, promoting basic hygiene care through education and actually providing access to the most vulnerable citizens.

It is important to clarify that this United Nations resolution does not mean however that the member states have to provide these services immediately to their entire population, because this is generally not feasible. They do not have to provide these services directly themselves, because they may do so through different stakeholders. They do not have to provide them for free, since they can and should be associated with a tariff.

These rights mean, to sum up, that the member states should define and implement appropriate, coherent and integrated public policies for water services, and also assume a major commitment regarding their implementation.

The Millennium Development Goals and the United Nations resolution regarding the access to safe drinking water and sanitation as essential human rights should not be seen as an end in themselves, but as opportunities for additional progress regarding universal access. Their implementation requires long-term planning, involves progressive realisation and will be highly financially and logistically demanding.

Even without being considered a human right, the services of solid waste management can benefit, through their similarity, from this new framework for the water services.

2.3 WHAT PUBLIC POLICIES FOR WATER AND WASTE SERVICES?

It should thus be the objective of all countries to promote the development of these services in order to improve the quality of life of local populations. These are however services with high investment and operational costs. It is essential to ensure that such development is sustainable, being carried out in a globally consistent manner, with proper coordination of all important aspects. This implies the existence of suitable public policies for the sectors that make use of a whole range of instruments of various kinds, globally competing for the same purpose and guaranteeing an optimisation of results in relation to the resources available. The implementation of appropriate public policies will avoid huge investments being made with no guarantee of obtaining the expected benefits for society.

It is thus a responsibility of governments to create the necessary conditions for the gradual generalisation of access of the entire population to these water and waste services, and conditions that must necessarily define appropriate public policies.

In general terms, any public policy concerning access to drinking water supply and waste water management, and similarly concerning access to solid waste management, should form a global and integrated, that is to say holistic, approach and include several components, which are described below:

- Adoption of strategic plans for the sectors
- Definition of the legislative framework
- Definition of the institutional framework
- Definition of the governance of the services
- Definition of the access targets and the quality of service goals
- Definition of the tariff policy
- Provision and management of the financial resources
- Construction of the infrastructure
- Improving the structural and operational efficiency
- Human resources capacity building
- Promotion of research and development
- Development of the economic activity
- Introduction of competition
- Protection, awareness and involvement of users
- Provision of information

The successful implementation of public policies for water and waste services depends on the ability to manage the implementation of all these components at relatively the same time, ensuring an effective global and integrated approach.

It is important to note that the implementation of one or a subset of these components is generally not sufficient to ensure the sustainability of the sectors, as this does not achieve the results expected in a long-lasting manner. For example, if it is possible in any region to make available and manage the financial resources to enable the construction of infrastructure, using funds from cooperation and development assistance, but without being provided with the other components, namely a good legislative framework, a suitable institutional framework, good models of governance of services, an effective tariff policy and human resources capacity building, then the odds of success are very small and the investments made will not have the expected return.

Moreover, it is necessary to take into account that the implementation of public policies for water and waste services should be progressive, due to their great complexity and the costs involved, focusing on the priorities of the country and giving special attention to the neediest users.

2.4 WHAT IS THE ROLE OF REGULATION IN PUBLIC POLICIES?

Regulation should be seen as a component of public policies on water and waste, one out of various, but which has a very important role given the fact that it promotes or controls most of the remaining components. It can be seen as the procedure of interpreting and implementing laws, policies and regulations, to achieve the global objectives for the water and waste services.

In most situations it is intended that the State is the promoter of the effective and efficient delivery of essential public services, at an appropriate level of quality, at social affordable prices and an acceptable level of risk for users, while simultaneously ensuring the economic, social and environmental sustainability of the providers of these services. The aim is to find the optimal balance between these goals, particularly to thereby ensure transparency towards users, regardless of whether the governance model is public or private.

When discussing the roles that may be assigned to the State, whether in a more minimalist view, focusing on essential functions, or in a more maximalist vision, including direct intervention in the economy, what emerges as a consensual aspect is the fact that regulation is one of the main functions of the State, albeit that the proposals for the model and level of intensity to be used in such regulation often diverge.

However, regulation is not the solution to all problems, but only a relevant instrument of public policy, and is more effective according to the level of consistency of that policy.

2.5 WHAT MUST BE THE REGULATORY APPROACH?

2.5.1 Regulatory model

Implementing a regulation model for the public drinking water supply, waste water management and solid waste management services should be carried out with much reflection, thus contributing to the improvement of all aspects regarding the services and not just in a partial manner, seeking in this way to find the ideal global solution.

When setting up a regulator for water and waste services, it is essential to define a clear and effective regulation model, which is rational with regard to the local context of these services, while also benefiting from the experience of regulation in other activity sectors and, of course, from international experience. This clarity is indispensable so that all the stakeholders involved in the sectors, especially the utilities, know the rules of the regulation model in advance and can decide their positioning with greater security.

The regulation model to be adopted may of course vary according to the services in question (water services, waste services or others) and with regard to the actual context in which the services will be developed, taking into account, in an integrated manner, technical, economic, legal, environmental, social and ethical aspects, and whether this is to be implemented within a short, medium, or long-term perspective, with stable rules independence, capacity, impartiality and transparency. It should be clear, simple and practical for the users.

The conceptualisation of the regulation model therefore depends on the existing context, that is, the real situation for the water and waste services and the surrounding political, economic, social and environmental context.

It is essential to understand that there are no universal solutions and, for each reality, whether this is a country or region, the most suitable regulatory model should be designed and then the details added. It is clearly a risk to adopt a regulatory model that has been imported and which has not been adapted to the reality of the country or region.

It is however very important that conceptualisation of the regulation model be carried out in an integrated and holistic manner, to take into account problems which are both specific and global in nature and seeking an integrated regulatory approach for the water and waste services, which can resolve the various problems separately but which can also find the optimal global solution, through a suitable balancing of the various aspects involved. The model should thus be applied through distinct components but also ensure there is close linking and in this way enhance the synergies between these same components. For example, economic regulation should be an enhancing factor but also benefit from quality of service regulation, and drinking water quality regulation should influence economic regulation and also benefit from it. This connection should be seen between practically all the components of the regulatory model.

In Portugal, over the last decade, ERSAR has designed and implemented a regulation model adapted to its national reality, specifying the corresponding procedures and creating the technological tools needed, which it has applied to around 500 utilities.

Based on this practical experience of around 12 years of regulation, a model of regulation is described based on an integrated approach (known in the abbreviated form as the model RITA-ERSAR) for water and waste services, seeking to find a suitable balance regarding the various aspects involved. This model aggregates the various strategic, technical, economic, environmental and social components, seeking an appropriate balance of the various perspectives at stake. It regulates both the sectors as a whole and also the utilities individually, and is simultaneously applied in a relatively similar manner to the different public services, specifically drinking water supply, waste water management and solid waste management, with suitable adaptations, seeking to find the optimal global solution.

This integrated regulatory approach for the water and waste services is implemented through two main levels of intervention, as described below:

- A first level, aimed generically at the sectors as a whole, designated as structural regulation of the sectors, involves a contribution towards the better organisation of the sectors, through the clarification of its operational rules, the drawing up and regular dissemination of information regarding the sectors, and capacity building and innovation for the sectors. The regulator is not focused on any utility in particular, but on the sectors as a whole, helping to create organisation, rules and tools for its good functioning. It therefore corresponds to macro regulatory intervention.
- The second level, designated as behavioural regulation of the utilities, consists of its legal and contractual monitoring throughout the life-cycle, economic regulation, quality of service regulation, drinking water quality regulation and user interface regulation. In contrast to structural regulation, the regulator is focused here on each of the utilities operating in these sectors. It therefore, in a supplementary manner to the previous level, corresponds to micro regulatory intervention, multiplied by the number of regulated utilities.

The need for effectiveness in regulating natural and legal monopolies leads to the supplementary utilisation of the structural regulation of the sectors and the behavioural regulation of the utilities. Disconnected utilisation is necessarily less effective than this supplementary utilisation.

In the case of the public drinking supply of water, waste water management and solid waste management services, as they remain relatively static through time, with gradual alteration of market and technological conditions, there tends to be a prevalence of behavioural regulation of the utilities over the structural regulation of the sectors. However, in periods with more pronounced revisions of public policies, there tends to be a reverse prevalence of structural regulation of the sectors over behavioural regulation of the utilities. The intervention of the regulator should of course be adapted to accompany such trends.

An analysis of the structural regulation of the sectors and the behavioural regulation of the utilities will now be carried out in more detail in the following text. Greater detail can be assessed in the publications (in Portuguese and English) referred to in the bibliographic references.

2.5.2 Structural regulation of the sectors

Structural regulation of the sectors is focused on the sectors as a whole and should contribute to its better organisation and to help to clarify its rules, such as restrictions on the entry of utilities into the market and functional separation measures, which define which entities or types of entity can provide these services.

Structural regulation also includes a set of measures to consolidate and modernise the sectors, both by making information available and enabling stakeholder capacity building. This regulation is a form of direct control over the surrounding and indirect context concerning the utilities, reducing or eliminating the possibility of undesirable behaviour. Within a preventative logic, it strongly conditions the form, content and nature of behavioural regulation, so much so that it needs to be supplemented.

Any definitions or alterations of public policies for the sectors should necessarily be accompanied by the regulator which, while of course not possessing the competency for any such definition, as these are decisions of a political nature, should, however, help to substantiate such choices, particularly in order to guarantee the protection of users' interests and safeguard the utilities' legitimate interests, as well as assessing the level of acceptable risk for society.

As regards the structural regulation of the sectors, the regulator should contribute towards:

* organisation of the sectors;
* legislation of the sectors;
* information of the sectors;
* capacity building of the sectors.

Each one of these structural regulation components will now be briefly described.

2.5.2.1 Regulatory contribution to the organisation of the sectors

In this structural regulation component, the regulator should contribute to the formulation of better public policies, towards their rationalisation and the resolution of any malfunctions regarding the regulated services and towards the organisation of the sectors, promoting for example an increase in the efficiency and effectiveness of the water and waste services and the search for economies of scale, scope and process.

It should afterwards monitor the national strategies adopted for the sectors, by accompanying their implementation and regularly reporting on any evolution or constraints.

2.5.2.2 Regulatory contribution to the legislation of the sectors

In this structural regulation component, the regulators should draw up proposals for new legislation or the alteration of existing legislation, for example at the level of the legal framework governing the systems, technical legislation regarding the water and waste services and the legislation governing regulation.

In this way it should contribute towards a clarification of the rules for the provision of these services through proposed legislation and the issuing of regulations and recommendations.

It should afterwards monitor the application of legislation in force and those regulations and recommendations, assessing their effectiveness and the need for any improvements or replacements.

2.5.2.3 Regulatory contribution to the information of the sectors

In this component of structural regulation, the regulator should make available and regularly disclose thorough and accessible information to all sectors' stakeholders, through the coordination and carrying out of the collection, validation, processing and disclosure of information regarding the sectors and the respective utilities, as well as making that information available with the subsequent increased public interest in consulting it.

It should thus contribute to consolidating an actual culture of concise and credible information which can be easily interpreted by all, extendable to all utilities, regardless of the forms of management adopted for the provision of such services. It should enable suitable knowledge to be made available, based on the information obtained from the numerous data created within the sectors, thus ensuring the fundamental right of access to information of all users and society in general.

2.5.2.4 Regulatory contribution to the capacity building of the sectors

In this component of structural regulation, the regulator should provide technical support to the utilities through the production of technical publications in partnership with knowledge centres, the direct and indirect promotion of seminars and conferences, support for events by third parties, conducting opinion studies and promoting research and development in the sectors, thus motivating the academic sector in this field. It should also make itself available to answer the various questions put to it by the different sectors' stakeholders.

In this way it can contribute to an improvement in the technical capacity building of the utilities and to encouraging the consolidation of the national business sectors.

2.5.3 Behavioural regulation of the utilities

At the same time, the strategy of the regulator should also include the behavioural regulation of the utilities in their actions in markets subject to regulation, with regard to the legal and contractual, economic, quality of service, drinking water quality and user interface aspects, which will be described below.

These regulation components are strengthened through performance assessment and benchmarking with the results of other similar utilities operating in different geographical areas. These mechanisms should adopt a logic which is pedagogic and appreciative of worth, benefiting the utility in some manner in terms of its performance regarding the average performance of all utilities. To achieve this, it is necessary that the regulator receives information from the utilities in the form of data so that previously defined performance indicators can be calculated and, after their validation, a comparative analysis is carried out with the historical records of the actual utility, so as to see the evolution through time of different management aspects, and compare with other similar utilities. This allows, in particular, performance levels to be defined and references to be established to enable the realistic setting of new efficiency targets. The results of this comparison should be made available publicly, in so far as this pressures the utilities in terms of their efficiency, as they naturally do not want to be seen in an unfavourable light, and so it implements a fundamental right available to all users.

As regards the behavioural regulation of the utilities providing water and waste services, the regulator should carry out:

- legal and contractual regulation;
- economic regulation;
- quality of service regulation;
- drinking water quality regulation;
- user interface regulation.

Each one of these behavioural regulation components will now be briefly described.

2.5.3.1 Legal and contractual regulation

In this component of behavioural regulation, the regulator should ensure the legal and contractual monitoring of the utilities throughout their life-cycle, specifically through analysing the tendering and contracting processes, contract modifications, contract terminations, and reconfigurations and mergers of systems, accompanying the carrying out of contracts and intervening where necessary in reconciliation activities between parties.

Legal and behavioural regulation should thus contribute to ensuring public interest and legality.

2.5.3.2 Economic regulation

In this component of behavioural regulation, the regulator should ensure the economic regulation of the utilities, thus promoting the regulation of prices to ensure efficient tariffs which are socially acceptable to users without prejudice to the necessary economic and financial sustainability of the utilities, within an environment of efficiency and effectiveness in the provision of their service. Economic regulation also includes the assessment of investments to be undertaken by the utilities.

As monopoly prices tend to be higher than those resulting from competing markets, obtaining lower prices that will allow the economic and financial viability of the utilities and will correspond to a fairer system for users requires major intervention from the regulator.

Economic regulation should thus contribute to promoting the economic and financial sustainability of those utilities, without prejudice to the economic accessibility to the services for the users.

2.5.3.3 *Quality of service regulation*

In this component of behavioural regulation, the regulator should ensure the regulation of the quality of service provided to users by the utilities, assessing their performance and comparing the utilities among themselves, through the application of a suitable selection of performance indicators, so as to promote effectiveness and efficiency, which represents an improvement in their levels of service.

Quality of service regulation is a way of regulating performances, which is inseparable from economic regulation. It constrains the permitted performances of the utilities regarding the quality of service they provide to users, steering the utilities in terms of effectiveness and efficiency, and embodying in this way a basic right of users.

This should therefore contribute to promoting an improvement in the levels of service provided to users.

2.5.3.4 *Drinking water quality regulation*

In this component of behavioural regulation, the regulator should ensure drinking water quality regulation, assessing the quality of water supplied to the users, comparing the utilities among themselves and monitoring any non-compliances in real-time.

As the quality of drinking water is an aspect of quality of service, and as it has a major interaction with economic regulation, there is a clear rationale for it to be regulated by the same body, although it is not absolutely necessary to follow this model and this is not the case in various countries.

It should thus contribute to promoting an improvement in water quality and public health.

2.5.3.5 *User interface regulation*

In this component of behavioural regulation, the regulator should ensure compliance by the utilities with consumer protection legislation and, in particular, undertake an analysis of any complaints and promote their resolution between users and the service provision utilities.

It should also foster the participation of service users, creating advisory mechanisms and disseminating information.

2.6 WHAT ARE THE CONCLUSIONS?

In summary, Figure 2.1 which follows is a graphic representation of the proposed RITA-ERSAR regulation model, which has been adopted by ERSAR, the Portuguese regulator, for more than a decade, corresponding to an integrated regulatory approach for the water and waste services.

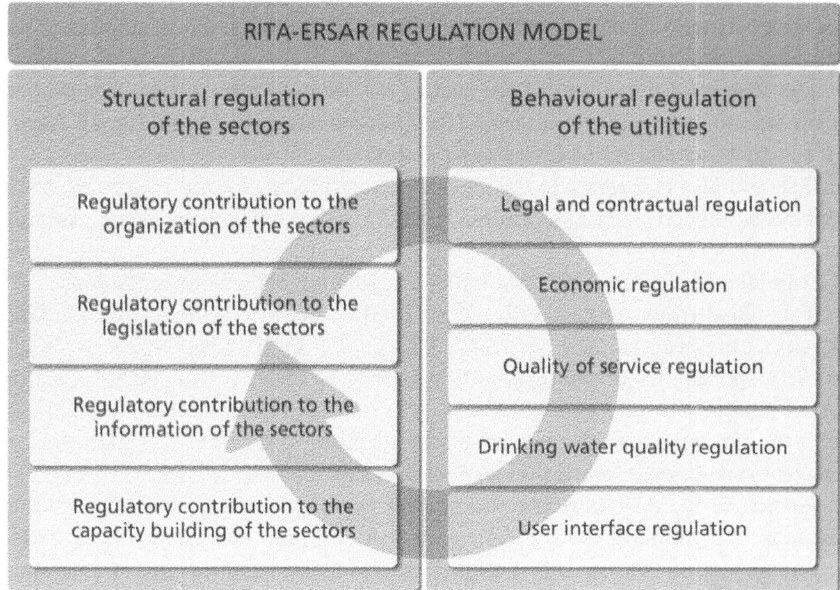

Figure 2.1. RITA-ERSAR regulation model.

As mentioned above, this model is based on two major levels of intervention: structural regulation of the sectors and behavioural regulation of the utilities. The components of the structural regulation of the sectors consist of contributions towards the organisation, legislation, information and capacity building of the sectors. The components of the behavioural regulation of the utilities consist of legal and contractual regulation, economic regulation, quality of service regulation, drinking water quality regulation and user interface regulation.

All these components must be perfectly linked to each other, so as to form a coherent and integrated model, the effectiveness of which depends on the synergies obtained between these components.

It is recommended that the regulatory authorities, materialised in one or more organisations, as major stakeholders in these sectors, ensure the implementation of regulatory models which form an integrated regulatory approach, for example for the public water supply, waste water management and solid waste management services, regulating both the sectors as a whole and the utilities individually, looking to find the optimal global solution for the populations.

In general, the model should include a level of structural regulation for the sectors, in which the regulatory authority should contribute to the organisation, legislation, information and capacity building of the sectors, and a level of behavioural regulation for the utilities, in which the regulatory authority should carry out legal and contractual regulation, economic regulation, regulation of the quality of service, regulation of the quality of drinking water and regulation of the interface with users.

The structural regulation of the sectors should contribute to the organisation of the sectors, particularly in terms of an optimal territorial organisation taking advantage of economies of scale, scope and process.

It should also contribute to the legislative instruments for the sectors, with the goal of helping to clarify its operating rules, an aspect which is of course essential for the proper delivery of these services through instruments with a variable degree of external effectiveness, specifically laws, regulation and recommendations.

It should also contribute to information for the sectors, developing and regularly disseminating accurate and accessible information to all stakeholders. It should be the responsibility of the regulatory authority to create a national information system and to consolidate a culture of concise, credible information which can be easily interpreted by all, extendable to all utilities, regardless of the forms of management adopted for the provision of services.

Finally, it should contribute to capacity building in the sectors, promoting research and development, creating innovation and endogenous knowledge and human resources capacity building with suitable technical and professional training to better carry out their activity, thus ensuring increased autonomy for the country's water and waste services.

The behavioural regulation of the utilities should play a role in the legal and contractual area, with the aim of ensuring that all of the stages of its life-cycle, from the design stages, any tendering process, contracting, service management, contract amendment and termination, are carried out in strict compliance with legislation and any existing contract, as is the case in situations involving delegation and concessions.

It should also play a role in the economic area, with the aim of ensuring the application of tariffs and a suitable tariff scheme, with appropriate behaviour and economic and financial efficiency from the utilities, promoting the rationalisation of prices to users at the same time as the economic and financial sustainability of the utility.

It should also play a role in the area of quality of service, with the aim of ensuring the provision of a suitable quality of service to users by utilities under the terms of its goals and the applicable law.

It should also play a role in the area of the quality of drinking water, with the goal of ensuring, in a continuous manner, the supply of an appropriate level of water quality by utilities, in accordance with applicable law, for the benefit of public health.

It should also play a role in the area of the interface with users, in order to better ensure the protection of users' rights through compliance with legislation for consumer protection, defence of the right to make a complaint and subsequently improving the quality of the relationship of the utilities with the users.

Regulatory authorities should act based on the principles of competence, exemption, impartiality, accountability and transparency.

If the political powers implement a suitable public policy that includes an integrated regulatory approach, if regulatory authorities and the utilities perform their activity appropriately and if users play their part in proactive citizenship, then the essential conditions have been met to promote the provision of public water and waste services with universal access, continuity, quality, efficiency and price equity, thus constituting an important factor for social and economic development.

2.7 BIBLIOGRAPHY

Baptista J. M. (2014a). Abordagem regulatória integrada para os serviços de águas e resíduos. ERSAR (in Portuguese).

Baptista J. M. (2014b). The regulation of water and waste services – an integrated approach. IWA (in English).

Chapter 3

Water regulation in the UK and its relevance to Spain

Michael Rouse
Oxford University, UK

3.1 INTRODUCTION

There has been a tendency to see regulation as part of privatisation, partly because of the wide interest in privatisation in the 1990s with the model adopted for England and Wales being closely examined by other countries. Rouse (2013) has long argued that regulation is just as relevant to other forms of water and wastewater service delivery. Independent regulation provides objectivity, transparency and greater public confidence. This doesn't necessarily mean having a body called a regulator because there are successful cases in which those important aspects are provided by other organisations such as water service associations and non-governmental organisations (NGOs). The historical attitude to regulation is illustrated by Figure 3.1 in which all forms of water service delivery are required to comply with general and environmental regulation, but with there being no objective economic regulation in the traditional municipal model, and increasing economic regulation as there is greater separation from government.

There is a strong case for "tilting" the diagonal line in Figure 3.1 to provide the benefits of regulation even at the municipal level, although in practice this may require the formation of "corporatised public" utilities to achieve the necessary separation and avoidance of political interference. A discussion on this aspect of successful governance is beyond the scope of this chapter.

The form of regulation adopted by a country is influenced by history. The statutory water companies in England, some of which were established over 150 years ago were originally controlled by rate of return limits and their stock paid a fixed return. It is appropriate to give an outline of the history of the water industry in England and Wales leading up to the privatisation regulatory regime introduced in 1989. Also the situations in Scotland and Northern Ireland are described briefly as these led to similar regulatory systems but with public sector ownership.

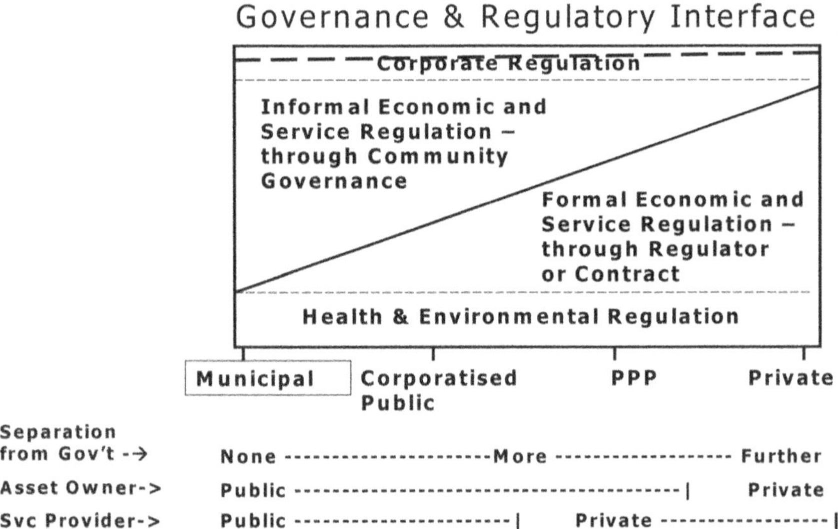

Figure 3.1 The four basic models of water service provider (Source: Paul Reiter).

3.2 HISTORY OF WATER INDUSTRY IN ENGLAND AND WALES UP TO 1989

The major changes began in 1974 although important steps had taken place from 1945 onwards. The Water Act 1945 began the process of consolidation in water supply with the requirement that local authorities form joint Water Boards. The impact of this was that the water supply bodies (over 1000) in existence at that time had consolidated to 198 in 1973. The Rivers Board Act 1948 was the beginning of pollution control through river management, and the River Boards became River Authorities in the Water Resources Act 1963, with the added responsibility of conserving water resources. The 1963 Act established a national body, the Water Resources Board to advise Government on long-term water resource needs. These were important steps but in the author's opinion the Water Act 1973 provided many of the key elements for the change to private sector ownership and management in 1989.

An authoritative account of the history is given in *The Development of the Water Industry in England and Wales* (2006). Here, an outline of the key changes is given. Prior to 1974 most of the water supply services, and all of the sewerage services, were managed by local authorities. There were 1620 public service water bodies in existence at that time; 198 local government organisations and Joint Water Boards, 1393 sewage and sewerage authorities, and 29 river authorities. Some of the 'Water Boards' were large like the Metropolitan Water Board serving London, which itself had been formed from 9 separate units in 1903, and there were some small local authority operations. As mentioned earlier there were also 33 statutory private water companies, some of which had been formed in the middle of the 19th century, which retained their historic private status and "regulation" until privatisation in 1989. The Water Act of 1973 made provision for the formation of 10 Regional Water Authorities (RWAs), structured on river basins with their boundaries being the watershed lines separating river systems. All of the Joint Water Boards and the local authority units, together with the river authorities, were incorporated into RWAs based on their locations within the river basins, although there were some anomalies between water and sewerage due to historic service provisions. At that time the Water Resources Board was abolished but another national body, the National Water Council was formed. There was great resistance from some of the local authorities to losing their utilities and legal action was taken against central Government to try to prevent the transfer of assets. These actions failed. Although there were clear benefits with a river basin approach the loss of direct linkage with city administrations resulted in separation of water and wastewater planning and city development planning. The changes also resulted in the loss of local identity and civic pride, which had produced some world leading operations at the time, and there are implications from this for cities of the future.

The need for action on sewage treatment had been highlighted earlier in *Taken for Granted*, the 1970 report of the *Working Party on Sewage Disposal* (1974), chaired by Lena Jeger, MP, set up by the ministry then responsible for water, the Ministry of Housing and Local Government. The "Jeger" report, and the poor quality inherited infrastructure discovered by the RWAs vindicated the decision to integrate many small utilities into viable sized units.

The argument for river basin authorities was not only viable utility size but also the need to manage river systems in an integrated way, both to optimise investment in water and sewage treatment and to achieve the desired environmental improvements in river water quality. The formation of RWAs in 1974 was heralded (Okun, 1977) as the way forward in water management. River basin management systems had been in place earlier elsewhere in Europe, one of the best known examples being the Ruhrverband (Ruhr River Association) which was created at the beginning of the 20th century, but the formation of the RWAs was the first time that river management, water supply and wastewater services had been integrated on a large national scale.

One other major change in 1974 was the cessation of government grants for infrastructure development, and from that time onwards water services had to be self-financing. In some areas, water service charges had been included in local taxation, so it was a new experience for many people to receive water bills. Until that time, consumers had not experienced the cost of their water services. It is not clear whether all local authorities reduced the level of local taxation by the appropriate amount and many people saw a significant increase in their total service payments. The formation, perhaps because people experienced an overall increase in "local" charges, was more controversial than the later privatisation.

Although the formation of the RWAs was regarded as a major step forward in the management of river systems, the RWA regulatory role came under criticism. The manufacturing industry was concerned that its water abstractions and discharges were being regulated by RWAs who themselves had a major impact on river quality due to sewage discharges. The term used to describe this at the time was that the RWAs were both "poachers and gamekeepers". Clearly there could be conflicts of interest and it is not desirable for an operator to also be a regulator. This anomaly was not addressed until the time of privatisation in 1989.

The Water Act 1989 established the following:

- The 10 RWAs, 9 in England, and 1 in Wales, were privatised through flotation on the stock market with shares being widely owned. Special provision was made for staff and customers of RWAs to obtain shares.
- The status of the then 26 Statutory Water Companies changed to plcs (public limited companies) and became regulated in the same way as the 10 ex-RWAs.
- The regulatory responsibility of the RWAs, together with the responsibility for flood control, was transferred to a newly formed body, the National Rivers Authority.
- Economic regulation became independent from Government and was handled by a new body, the Office for Water Services, Ofwat.
- Powers were given to the Secretary of State to make regulations to define the "wholesomeness" of drinking water. This resulted in the Water Supply (Water Quality) Regulations 1989.
- A Drinking Water Inspectorate, based within the Department of the Environment, was formed to enforce drinking water quality regulations.
- Customer Service Committees were established with regional offices.

The full provisions can be seen in the 1989 Act and were later part of the consolidation of water legislation in the Water Industry Act (1991) which was "An act to consolidate enactments to the supply of water and the provision of sewerage services, with amendments to give effect to the recommendations of the Law Commission". According to the allocation of powers in the Act, both the Secretary of State (i.e. the Government Minister responsible for Water) and the Director General, Ofwat, could issue an enforcement order if a water company contravened any condition of its appointment licence or any statutory requirement. The powers for action in Wales were transferred to the Welsh Assembly on its establishment in 1999. The National Rivers Authority became part of the Environment Agency (EA) in 1995. The powers on enforcement of drinking water quality were transferred to the Chief Inspector of the Drinking Water Inspectorate (DWI) in 2003. The regulatory structure established at that time is essentially the same at the time of writing and is shown in Figure 3.2. The government department now responsible for policy in England is the Department of Environment, Food and Rural Affairs (Defra). Through amalgamations and takeovers, the number of water service providers has reduced from 36 in 1989 to 23 in 2014. The roles and responsibilities are discussed later in the chapter.

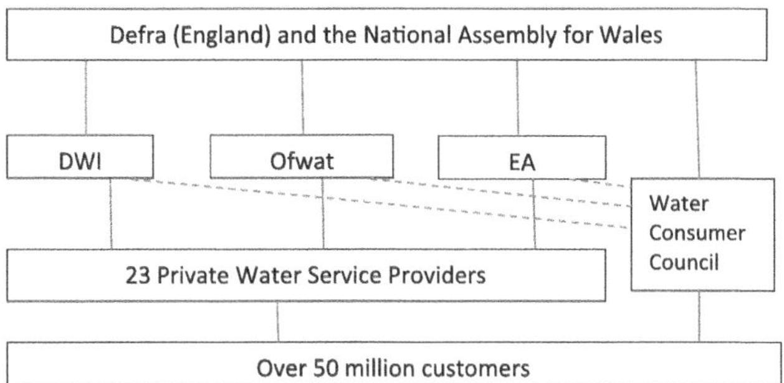

Figure 3.2 Regulatory structure of water and wastewater in England and Wales.

The water company in Wales, Welsh Water (Dwr Cymru), was initially an equity-based company, the same as the English companies. In 2001 to avoid a takeover by an American energy company, some influential people in Wales raised capital to buy the company and converted it into a "mutual" in which it does not have ownership but trustees.

3.3 SCOTLAND AND NORTHERN IRELAND

Prior to 1975 water and wastewater service providers were largely municipal. In 1975, as part of local government reform, water and wastewater operations were incorporated into nine Regional Councils. In 1996 water and wastewater services were separated with the formation of three Water Authorities, namely East, West and North of Scotland. These new authorities were regulated by a new body, the Water Industry Commissioner for Scotland, which was modelled on Ofwat. A drinking water inspectorate was established within the Scottish Department. Six years later the three authorities were amalgamated to form one organisation, Scottish Water, a corporatised public company.

There were associated changes to water environmental management with the formation of River Purification Boards in 1973 with some integration to have seven such boards in 1989. The 1995 Environment Protection Act created the Scottish Environment Protection Agency (SEPA) in 1995. Following a referendum in 1997 there was devolution of powers to Scotland through the Scotland Act 1998 and the establishment of the Scottish Parliament.

There have never been any private operations in Scotland, but a similar regulatory model to that of England and Wales has been adopted with Scottish Water coming under the regulatory structure shown in Figure 3.3. However, to deliver a major investment programme, which included addressing the backlog of infrastructure refurbishment, Scottish Water entered into partnering arrangements with the private sector (*Investing in Scotland's Water Industry: Improvements Delivered in 2006–10*).

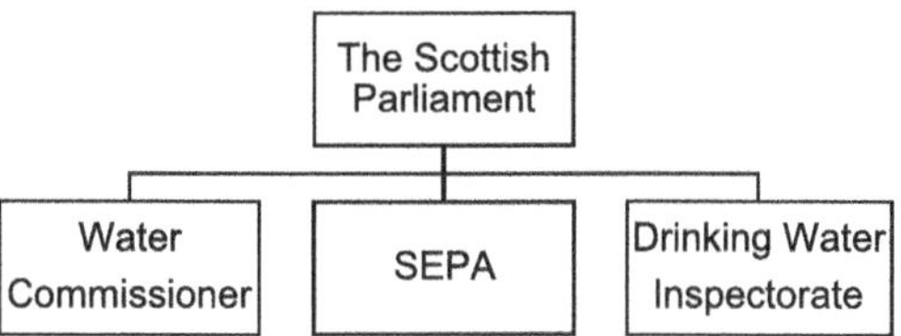

Figure 3.3 Water and wastewater regulatory structure in Scotland.

In Northern Ireland through the Water and Sewerage Services (Northern Ireland) Order 2006 the provision of these services transferred from the Water Service, an agency of the Department of Regional Development (DRD) to a government owned company, Northern Ireland Water Ltd (NI Water). This came into effect on 1 April 2007 when a water directorate was formed within the utility regulator of Northern Ireland. A regulatory approach similar to that of Scotland is being adopted.

3.4 THE DEVELOPMENT OF REGULATION IN ENGLAND AND WALES

At the time of privatisation in 1989 there were major water and wastewater challenges. The old infrastructure had been neglected due to government restrictions on investment. There were many treatment plants, both water and wastewater, requiring refurbishment and upgrading, and in particular there was a backlog of investment in the refurbishment of underground sewers and water mains. At the same time major investment was required to meet drinking water quality regulations and to improve the quality of rivers and coastal waters. Privatisation demanded transparency on these requirements so that the financial markets were able to assess the investment requirements and risks. An open objective means of determining investment programmes and associated water charges (bearing in mind that there were no government subsidies) became essential. This required strong and independent regulation, not only on the economics but also on quality. There also had to be an effective consumer voice. The regulatory structure shown in Figure 3.2 was established to meet those requirements. Each of the functions is discussed in turn.

3.5 OFFICE OF WATER SERVICES (OFWAT)

It is valuable to first consider how Ofwat operated at the beginning, and then to discuss how it is evolving leading to some significant changes which will come into effect as a result of a Water Bill going through parliament at the time of writing.

Ofwat's role is to ensure that the water companies give consumers a good-quality, efficient service at a fair price. Ofwat sets price limits, promotes economy and efficiency, protects consumers, operates a guaranteed standards scheme and sets and audits leakage targets. Ofwat also compares company performance using performance measures and benchmarking (Service and Delivery: Performance of the Water Companies in England and Wales 2009–10). Ofwat has to protect consumers' interests, but at the same time ensure that the water companies are able to finance the required improvement programmes.

The main function of water and wastewater service price setting is best illustrated through describing the five-year periodic review planning process. The steps in the process are:

- Ofwat, the economic regulator, publishes the timetable of events.
- Customers are consulted through various market research exercises, some at the beginning and some later when Ofwat has produced its interim determinations. The surveys are organised jointly by Government, the regulators, the Consumer Council for Water, and the water companies and by the individual stakeholders, depending on the scope of the surveys. The aim is to gain a better understanding of which improvements customers would like to see and how much extra, if any, they are prepared to pay for them.
- Government publishes its objectives including, in outline, the service improvements it would like to see achieved during that period.
- The DWI (Drinking Water Inspectorate) and the EA (Environment Agency) then prepare guidance for the water companies on what is required to achieve any desired quality improvements.
- On water quality-related aspects, the water companies then liaise with the DWI and the EA on how they will meet the requirements, and the two quality regulators say whether the proposals are appropriate.
- The water companies then submit business plans giving the improvement schemes and their costs.
- Ofwat considers these plans and whether the costs are realistic and, taking into account both the costs and the required efficiency improvements, calculates the required water prices to pay for the work whilst giving the companies a reasonable profit. These are given as price limits, called price caps, which are the maximum average water charges allowed to be applied by water companies.
- Reporters assist Ofwat in the submission evaluation process. Reporters are independent professionals who, although appointed by the water companies, have a primary duty of care to Ofwat. Their important role is discussed later.

- These price limits are then reviewed, including by Government who consider whether the quality and other service improvements are affordable or whether some aspects should wait until a later period.
- There are then discussions between Ofwat and each company before Ofwat announces its decision on the price limits for each company.
- Should a company wish to challenge the "settlement" it can appeal to the Competition Commission which carries out a completely independent period review, and whose decision is final.

This process is carried out with all the documents in the public domain. The process takes around 18 months but has the advantage that all the stakeholders are involved in an integrated way.

An incentive-based approach is taken in which water companies have an incentive to outperform. Outperformance benefits accrue to the companies during the five-year period but the price-cap starting point for the next five-year period is at the higher efficiency level which then benefits the customers. The formula used by Ofwat, which is based on a consumer price index minus inflation approach, is $K = RPI - X - Po + Q + S + V$, where

- K is the cap on annual increases in tariff basket
- For each company there is an allowable K for each year
- RPI is the retail price index as the measure of inflation
- X is the efficiency factor which is different for each company
- Po is the adjustment from the previous 5 years based on outperformance
- Q is the cost of capital for quality improvements
- S is the cost of service improvements
- V is the cost of providing additional supply volume

Notes on formula:

(1) The companies consult with their local Consumer Council on the tariff basket.
(2) Ofwat determines the X efficiency factor for each company based on comparative performance (comparative competition). Those companies that are shown to be more efficient have lower X numbers.
(3) Po is the difference between the K allowed in the previous five-year determination and what a company achieved.
(4) Quality improvements cover both drinking water quality and environmental water quality improvements.
(5) Service improvements cover for example times that a system is out of service for repairs. There is one item in a guaranteed standards scheme which is discussed later.
(6) In all improvements involving capital investment the allowed cost of capital is a critical element and probably the most controversial, see below.

Customer service performance is regulated by Ofwat through a Guaranteed Standards Scheme (GSS). The guaranteed standards of service are laid down by the Government. The scheme sets out the standards and the conditions under which water and sewerage customers are entitled to compensation. The scheme includes the following:

- Making and keeping appointments;
- Responding to account queries;
- Responding to complaints;
- Interruptions to supply, both planned and unplanned;
- Flooding from sewers which includes a minimum payment for flooding of property;
- Low pressure of water supply.

An example of a compensation payment is £25 minimum for low pressure.

3.6 SOME KEY FACTORS IN ECONOMIC REGULATION

3.6.1 Financing infrastructure

Water is a very capital intensive business. The infrastructure, particularly the underground water mains and sewers, make up a major part of the cost of the water service. However, compared with most businesses, asset lives are long, and outside the range of normal accounting depreciation practice. Partly for this reason, there has been a tendency to ignore the need for capital replacement. Most countries are now faced with the challenge of not only financing new infrastructure for system expansion and quality improvements, but also meeting the large investment required in refurbishing existing systems. In a regulatory environment, whether public or private, where water companies are required to be self-financing the cost of capital is critical in the tariff calculations. For sustainability, depreciation allowances have to be appropriate to fund ongoing replacement and refurbishment. In England and Wales, the cost of capital is a "number" hotly contested between Ofwat and the water companies. If water companies can obtain finance, either through equity, or more commonly from borrowing, at a rate which is lower than that set in the regulatory process, then water companies can increase profits through capital investment. This encourages capital investment, but the customers may pay more for their service in the short-term even though their bills may be lower at the next review. However, if the companies are unable to obtain capital at or below that set in the regulatory process, there is a risk that water companies will be unable to meet the investment requirements. The challenge for the regulator is to get the level "right" such that water companies are more likely to take the better option between a capital and a revenue solution. As the process of the periodic review begins to set the price caps for the period 2015–2020, Ofwat has issued a discussion paper (Setting price controls for 2015–2020, 2014) on its intentions on financing with a lower figure for cost of capital. This has resulted in reactions from the water companies as illustrated in the financial press (Financial Times, 2014) and (Daily Telegraph, 2014). The final agreed rate will emerge on completion of the periodic review.

The impact of capital costs on water charges is illustrated by Figure 3.4 which shows that the sum of the capital-related elements is 58% of the water bills. This shows that the financial conditions are likely to have a much bigger impact on the regulatory approaches than operational efficiency factors, although −X is an important part of the price cap formula.

Components of the revenue requirement for 2010-15

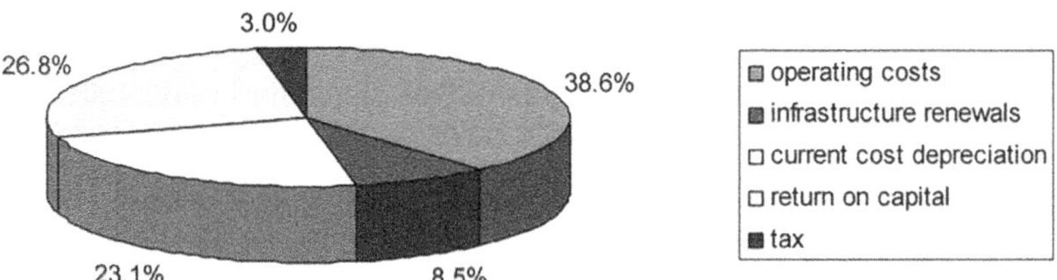

Figure 3.4 Cost components in water bills (Source: Ofwat discussion paper on financeability and financing the asset base, march 2011).

3.6.2 Determining the efficiency factors

Ofwat uses comparative competition measures (PR09/39 Ofwat 2009) to drive efficiency improvements. The X in the price cap formula is made up of two components, an overall improvement factor applying to all companies, and a catch-up factor applying to those companies who are calculated as less efficient. Ofwat uses econometric and unit cost models to assess relative efficiency. Water companies are assessed in five bands, A-E, A being the most efficient, and X is determined accordingly. The process relies on receiving accurate information form the companies.

3.6.3 Checking information

Until "recently" Ofwat used Reporters to certify and report their opinion on the regulatory information companies submitted. These appointees were paid for by the water companies with a duty of care to Ofwat. The certification process required Reporters to work alongside the companies, checking that company material assumptions were exposed in their submissions, and that Ofwat policy guidance and information definitions were followed by the companies. In giving evidence (*Minutes of Evidence*) to a House of Commons Select Committee on Environmental Audit in 2000, Ofwat referred to a review it commissioned on the effectiveness of the reports which concluded that the "Reporters" process is valuable to the Director (Note that at that time regulation was the responsibility of an individual, the Director General) in giving him an objective evaluation of the soundness and validity of the information employed by the water companies in the development and presentation of their business plans. It concluded the Director should feel confident in relying upon this evaluation.

The Scottish regulator also relies on Reporters to check data validity. In a report for the Water Commissioner, consultants said "Reliance was placed on the Reporter's role in challenging Scottish Water's submissions, confirming the basis of the Standard Costs presented and commenting on any material concerns such as data sources, methodology, assumptions and compliance with definitions."

The use of Reporters came into focus in relation to Government policy for de-regulation to reduce the regulatory burden on industry. A review of Ofwat (Review of Ofwat and Consumer Representation in the Water Sector, 2011) was commissioned by the Government. Water companies included the cost of Reporters and their audits of data as one of a list of items incurring (in their view) unnecessary cost. As a result, Ofwat has stopped using Reporters, although some water companies find them useful and will retain their services. The possible implication of this de-regulatory step is discussed later in the chapter.

3.7 THE DRINKING WATER INSPECTORATE

The Drinking Water Inspectorate (DWI) was set up at the time of privatisation of the Regional Water Authorities. The DWI is independent with enforcement and prosecution powers. Its work includes:

- Checks on adequacy of risk assessment related to water safety planning;
- Making inspections of water treatment plant, service reservoirs and laboratories to check compliance with regulations;
- Carrying out audits of the self-monitoring of drinking water quality by water companies for compliance with regulations;
- Taking enforcement action for non-compliance with regulations;
- Prosecuting for the supply of water 'unfit' for human consumption (this is discussed later in the chapter);
- The preparation and issue of annual independent reports;
- Pursuing consumers' complaints related to drinking water quality working closely with the Consumer Council for Water;
- Providing an enquiry service for consumers on all aspects of drinking water quality.

The DWI plays a major role in the periodic planning process in interpreting government objectives in delivery terms to provide guidance to water companies in the preparation of their business plans. The DWI reviews water company proposals to assess their appropriateness to meet the requirements, and monitors the legally binding delivery programmes.

The success of drinking water quality regulation is shown in Figures 3.5 and 3.6 and Table 3.1. Figure 3.5 shows the overall compliance based on around 3 million tests per year for the period up to 2003. The regulations changed in 2004 and the ongoing level of compliance with a different base is shown in Table 3.1. There are a few, but different cases, of non-compliance each year suggesting that an achievable maximum has been reached and maintained.

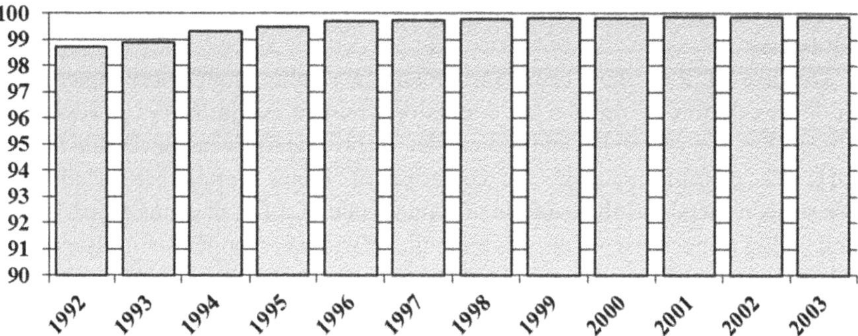

Figure 3.5 Percentage of tests complying with drinking water quality standards in England and Wales over the period 1992–2003. Source: Drinking water inspectorate annual reports: www.dwi.gov.uk.

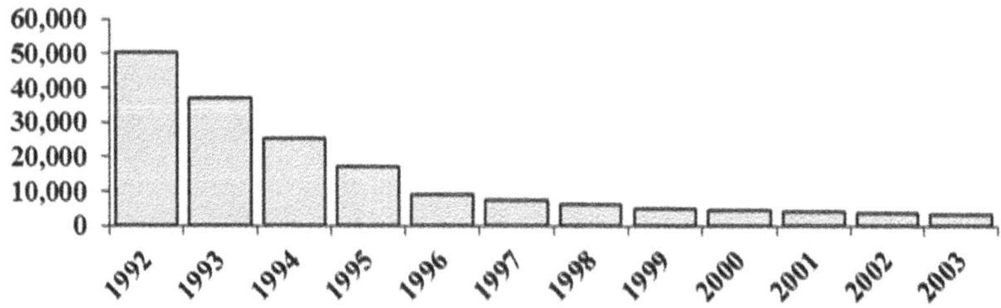

Figure 3.6 Number of tests breaching the drinking water quality standards over the period 1992–2003. Source: Drinking water inspectorate annual reports; www.dwi.gov.uk.

Table 3.1 Drinking water quality compliance with regulations.

Year	2004	2005	2006	2007	2008	2009	2010	2010	2011	2012
Compliance %	99.4	99.6	99.6	99.6	99.6	99.5	99.6	99.6	99.6	99.6

Source: Drinking Water Inspectorate Annual Reports: www.dwi.co.uk.

For each non-compliance the DWI considers whether it is "trivial" (not defined and considered for each case) and unlikely to reoccur, in which case no action is taken, or whether enforcement action should be taken. Enforcement decisions are legally binding. If the non-compliance results from circumstances outside the control of a water company, the cost of the necessary remedial actions may be included in costs for inclusion in the next periodic review, or if it were to make a substantial difference to a company's overall costs, Ofwat may make an adjustment ahead of the next review.

As already mentioned, in addition to enforcement powers, the DWI can prosecute a company for supplying water "unfit for human consumption". Water unfit for human consumption is not defined in legislation, and it is for a Court to decide on a case-by-case basis. In practice, it can be either water that results in illness, or water that is unpalatable for aesthetic reasons of taste, smell or appearance. For example, very rusty water has been regarded in Court as unfit for human consumption. In their defence water companies have to show that they "had no reasonable grounds for suspecting that the water would be used for human consumption" or that they "took all reasonable steps and exercised all due diligence" for securing that the water was fit for human consumption on leaving its pipes. Such prosecutions by the Drinking Water Inspectorate are infrequent. The total number of prosecution cases can be numbered in tens, whereas enforcement actions can be measured in thousands. The enforcement process enables more rapid improvements, to the benefit of water consumers. The prosecution powers are available as an important deterrent, with Courts able to impose a fine against a company or an individual or to give a prison sentence of up to 2 years. Together the enforcement and prosecution powers provide for a very effective regulatory regime to protect consumers.

3.7.1 The environment agency (EA)

(The EA covering England and Wales is discussed. The agencies in Scotland and Northern Ireland operate in a similar way).

The EA is a large organisation with environmental protection responsibility across the whole spectrum of environmental matters. It is also responsible for river navigation and flood management. On the latter it was under a lot of pressure during the 2013–2014 winter floods in England during which those affected by floods suggested that the EA had given greater priority to ecological considerations than to flood management. In particular, it was criticised for not having carried out river dredging. Political pressure has required this practice to be restarted. The EA has a similar difficult task in balancing its environment protection and water resource responsibilities, and was criticised in a House of Lords Committee report (Water Management, 2006), following the 2003 "drought", as having given too little attention to water resource provision. The author is not being critical of the EA but is pointing out the difficult task the EA has in achieving the right balance across conflicting responsibilities, with associated political dimensions.

The EA says it was established to protect and improve the environment and to contribute to sustainable development. It lists its responsibilities as:

- Regulation of major industry;
- Flood and coastal risk management;
- Water quality and resources;
- Waste regulation;
- Climate change;
- Fisheries;
- Contaminated land;
- Conservation and ecology;
- Navigation.

In dealing with its joint responsibilities of water resource management (given in the list above under water quality and resources) and environment protection, the EA has a number of tools, including those of approval of water company water resource plans, abstraction licences and discharge consents. The big water picture relates to river management, including responsibility for the implementation of the EU Water Framework Directive. Water companies are required annually to submit updated 25-year water resource plans to the EA and these plans provide the basis for water resource aspects in the periodic reviews. In considering the plans, the EA has to consider both water resource needs and the impacts of that abstraction on the environment.

Any abstraction from lakes, rivers or aquifers requires a licence from the EA, other than for those having individual wells not used for commercial purposes. Many of the major abstraction licences had been granted to water companies in the past at a time when there was less concern over the impact of low river flows. In legislation currently going through parliament, the EA will have powers to "renegotiate" these abstraction licences.

All wastewater treatment works are required to have a permit to discharge to the environment in the form of a consent from the EA. The consent has conditions that are designed to protect the receiving watercourse. The larger the discharge and the smaller the watercourse the more stringent the conditions are likely to be. In addition, the European Urban Waste Water Treatment Directive requires a minimum of secondary treatment for all works serving a population above 2000. Secondary treatment (settling and biological treatment to reduce the organic matter content) typically removes 95% of BOD (Biochemical oxygen demand), 95% of suspended solids, 29% of nitrogen and 35% of phosphorus from the untreated sewage. The procedure for obtaining a consent is given on the EA website (http://www.environment-agency.gov.uk/business/topics/water/32038.aspx).

The EA monitors river water quality. Until 2010 this was through a monitoring system called the General Quality Assessment Scheme (GQA). Environmental regulatory action has produced significant improvements, as can be seen in

Figure 3.7 which shows the percentage of rivers in England and Wales achieving either good or excellent quality status.

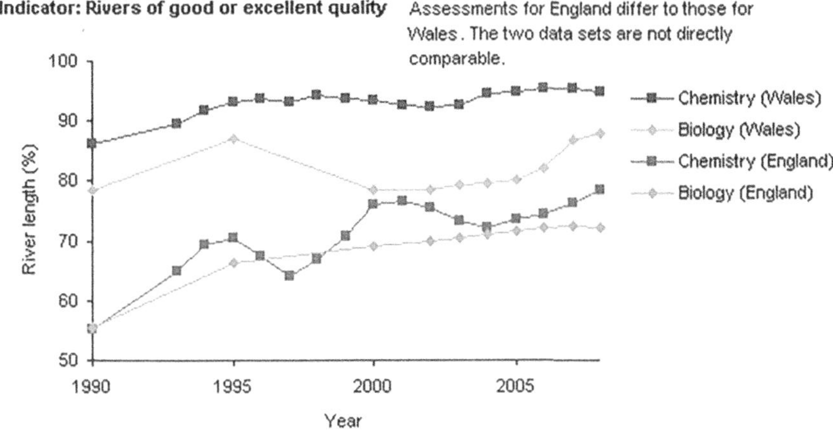

Figure 3.7 Percentage of rivers in England and Wales achieving good or excellent quality (Source: Environment agency (England and Wales): www.environment-agency.gov.uk).

The GQA had six classifications, very good, good, fairly good, acceptable, fair and poor, based on twice-a-year biological surveys and 12 samples per year for chemicals and nutrients. The results showing a comparison of the chemical results for rivers and canals combined with the good categories added together between 1990 and 2004 is given in Table 3.2.

Table 3.2 Comparison of chemical condition of rivers and canals in England and Wales between 1990 and 2004.

Chemical Condition	England		Wales	
	1990 (%)	2004 (%)	1990 (%)	2004 (%)
Good	43	62	86	94
Fair	40	31	11	4
Poor	14	7	2	2
Bad	3	1	1	0

Source: Environment Facts and Figures, Environment Agency.

With the introduction of the European Water Framework Directive (WFD), a new approach was introduced in 2008. To establish continuity of comparison,

GQA was run in parallel until 2011. WFD monitoring is more stringent and uses a risk-based approach for classification. It uses a principle of "one out all out" which means that classification is based on the poorest individual result. It uses 30 measures covering ecological status and chemical status. It covers not only rivers but also estuaries, coastal waters, lakes and groundwater. All results are published.

3.8 WATER CONSUMER COUNCIL

A consumer "watervoice" is not a regulator but is an important part of a regulatory system, particularly in a UK type system in which water and wastewater service providers are not municipal and there are no elected officials involved. The Water Consumer Council has an important role in providing an official channel for consumer input to periodic reviews and in the handling of customer complaints. For a description of the Water Consumer Council role and work see reference (Rous, 2013, pp. 234–237).

3.9 EXPERIENCE OF A REGULATED WATER AND WASTEWATER SERVICE SYSTEM SINCE ITS INTRODUCTION IN 1990

At the time of privatisation there was a backlog of investment both in ageing infrastructure and in meeting drinking water and environmental quality standards. Between 1990 and 2012 around £100 billion was invested in water and wastewater in England and Wales, twice the rate during the preceding public sector period. This has, or is, being paid for through water charges as there are no government investment subsidies. During the period 1990 to 2007 average charges increased by 37% in real terms. Ofwat estimates (Water UK Press Release, 2007) that efficiency improvements have absorbed 70% of that investment cost without which the increase in charges would have been much greater. However, the increases in water charges are uneven with some areas, such as the South West of England, where charges have become a more significant part of domestic expenditure. This is due to major investment in long distribution systems and on long coastlines falling on a relatively small population. The author believes that an opportunity was missed in the past when larger water companies were prevented from taking over South West Water, as costs would have then been spread over a bigger population. Government has introduced subsidies to those individual consumers in the area.

Drinking water quality has improved to virtually 100% compliance with regulations based on the requirements of the European Drinking Water Directive (98/83/EC) – see Figures 3.5 and 6 and Table 3.1. There have been improvements in river quality, partly due to improved sewage treatment discharges, although there are many more factors associated with environmental water quality. There has been major investment on refurbishing deteriorated water mains and sewers. Water distribution leakage has reduced but there is still much to be done. To put that challenge in the context of logistics, the water mains of England and Wales, if put end to end, would go round the world six times.

There can be many reasons given for success; during a long period of evolution up to 1989, when water was already self-financing with full-cost recovery from charges, investment levels became free from government public borrowing constraints, there was a strong regulation system in which water charge decisions could be taken based on need uninhibited by political considerations, good levels of private sector commercial energy, and good knowledge within the industry of problems and how to solve them. In the author's view all those things contribute to success but, significantly, in the early years there were experienced and knowledgeable water "people", with a public service ethic, who were able to make progress in a more dynamic working environment. In other words, there was a blend between public service values and private sector energy. There are now very few senior water managers remaining who grew up in the public sector. It is argued that this is a good thing as competition will ensure high standards. That is true in the long term as bad businesses fail. Generally that doesn't matter, as in most business areas failure doesn't result in public health consequences. It does not make sense for public health to be put at risk for short-term gains (i.e. marginally lower water bills). In a wholly business-driven environment regulators have a critical role in protecting public health. This aspect is returned to in a discussion of the Water Bill going through Parliament at the time of writing. It might have been expected that, due to major changes of 1974 in introducing catchment-based water and wastewater services and the introduction of privatisation in 1989, there would be no further radical changes, but water is never far from politics.

3.10 WATER BILL 2014

One of the objectives of the Coalition Government of 2010–2015 has been to reduce the "burden" of regulation, especially to cut "red tape" for small businesses, but all industrial regulation has come under scrutiny. Three government-commissioned reports, named after their authors, have had an influence on recent water legislation; in time sequence The Walker (The Independent Review of Charging for Household Water and Sewerage Services, 2009) and Gray (Review of Ofwat and Consumer Representation in the Water Sector, 2011) Reports (2009) and the Cave (Independent Review of Competition and Innovation in Water Markets, 2009) Report (2011).

The Walker Report, the Independent Review of Charging for Household Water and Sewerage Services, although best known for its recommendations on domestic metering, covered a number of other important aspects. It made recommendations on water efficiency, the problem of bad debt, flooding and surface drainage, high water charges in the South West of England, and a package of help targeted on low-income customers. In the Water Industry (Finance Assistance) Act 2012 provision was made for customers in the South West of England to receive a £50 reduction in their bills commencing in April 2013. Section 44 of the Flood and Water Management Act 2010 enables water and sewerage companies in England and Wales to include social tariffs in their charges schemes, reducing charges for households who would otherwise have difficulty in paying their bill in full. The Act requires the Secretary of State to issue guidance to water and sewerage companies and Ofwat, and requires companies and Ofwat to have regard to the guidance. A private members Bill in 2013 (i.e. not introduced by Government) aimed to limit water bills to 5% of disposable income, but there was insufficient parliamentary time for it to become an Act, however under the provisions of the 2010 Act all water companies are introducing social tariffs.

The Cave Report, the Independent Review of Competition and Innovation in Water Markets, recommended a much greater market approach to water services. The report covered water abstraction, discharge consents, competition and research. On abstraction it recommended fully tradable licences and a more risk-based approach to the allocation of licences. It recommended that discharge consents should also be tradable but subject to environmental impact assessments. A major recommended change was to move from vertical integration in which water companies are responsible for water resources, water treatment, water distribution and retail services to individual licences for each component as exist in the energy industries. The report recommended legal separation of water companies' retail functions. Government issued a white paper (Water Bill, 2013–14) in December 2011 in which it stated that it was not proposing a move away from vertical integration saying "it did not wish to take risks with a successful model given the challenges ahead". It changed its mind in the drafting of the Water Bill 2013–2014. Cave was very critical of the low level of research being undertaken by the water companies, saying that "The UK and Welsh Assembly Governments, the industry, regulators, suppliers, the research councils and the Technology Strategy Board and other stakeholders should come together to form a national water research and development body and agree a shared research and development vision for the industry." It suggested an annual £20 million research fund for 10 years with Ofwat being given a statutory duty to promote innovation. There has been no reference to research in subsequent legislation.

The Gray Report was a review of Ofwat and consumer representation in the water sector. On consumer representation the Consumer Council for Water (CCWater) was given a clean bill of health with reference to it having an important role in the introduction of social tariffs. A statement in the report read as follows "The review team is strongly of the view that effective consumer representation will be required in the sector for the foreseeable future and that the functions currently undertaken by CCWater should be maintained and protected in any new approach. Indeed, we can see areas in which the consumer role could be extended." On the review of Ofwat three key comments related to the periodic review were as follows:

- Ofwat needs to engage more constructively and effectively with the full range of stakeholders in the sector and be more transparent in its decision making.
- Ofwat should set clear targets and timescales for a reduction in the burden of the price control and compliance processes and enter into a joint project with the industry to achieve this.
- Ofwat should seek to ensure that the future framework of incentives provides the right balance between rewards and penalties.

The second one of these has had important implications. The water companies have complained about the administrative burden associated with the data returns, including during the periodic review and the role played by the Reporters. To comply with the requirement to reduce the regulatory burden, Ofwat has ceased to require the use of Reporters. There are arguments for and against this decision. The reduction in burden is a clear benefit (although the impact on consumers' bills is probably negligible) but what is the impact of Ofwat no longer having direct information on the true costs? It is argued that the introduction of more competition will mean that the market will take care of prices. In this context there has been criticism of the large energy companies dominating the electricity and gas markets resulting in high prices, but with a regulator who has insufficient knowledge of true costs. The author doesn't know whether or not this is true, but it must be an area of concern.

At the timing of writing the Water Bill 2014 has almost completed its passage through Parliament, and it can be expected to be law later in the year with implementation of the competition measures in 2017. The Welsh Government has decided not to introduce the competition measures so those aspects relate only to England. In introducing the Bill in 2012 the Secretary of State said:

- This draft Bill will create a modern customer focused water industry and for the first time all businesses and other organisations will be able to shop around for their water and sewerage suppliers.
- By slashing red tape we will also stimulate a market for new water resources and incentivise more water recycling.
- This will ensure that the water industry continues to provide an affordable and clean water supply which is essential for the nation's economic growth while at the same time protecting the environment for future generations.
- Businesses, charities and public organisations with multiple sites will also be able to receive just one combined water and sewerage bill for all their offices and buildings across England and Scotland.

Beyond political speak what are the main provisions in the Bill (2013–14)? Some comments relate to the author's interpretation of the provisions.

- The unbundling of the combined licence with individual licences for what are called retail and upstream services. (Note: New suppliers ("licensees") could provide either retail services alone, or both retail and upstream services. Retail services are defined as "customer-facing services, for example billing, meter reading and call centre services". Water would still be physically supplied by the incumbent company; the licensee would pay the incumbent company for the water and for use of its network to deliver it to the customers. Upstream services are defined as "the elements of the water and sewerage value chain that do not directly involve the customer" and could include activities such as the abstraction, storage, treatment and distribution of water. Upstream sewerage services would include the collection, treatment and disposal of wastewater and sewage. The intention is that upstream competition would create a "market for the sale of treated or untreated water into the supply system or the disposal of waste from the sewerage system". Services traded within this market could include developing a new water source and selling water from it to an existing water company, or developing a more environmentally friendly way of treating wastewater, re-using it for industrial use or disposing of sewage sludge. It is stated that "upstream competition will also make it easier for water companies to buy and sell water from each other". The market will only extend to some upstream services—the original incumbent water companies will still own and manage the network pipes that transport the water, wastewater and sewage and will continue to provide all the distribution services.)
- Access (to networks) prices to be determined by Ofwat based on costs and a common methodology
- Allowing retail infrastructures (mains, pipes, storage and treatment) to be connected to the existing primary networks. However, the main networks, because of logistics and cost, would be monopolies.
- All non-household customers, whatever the volume of water consumed being eligible for competing suppliers. However, this does not apply to domestic customers.
- As part of meeting concerns on risk of potential lack of competence of newcomers, a requirement for the quality regulators to be consulted on new entrant company applications.
- The Secretary of State and Welsh Ministers to issue and revise high-level guidance relating to Ofwat's charging rules—both charges to consumers, and charges between incumbent water companies and licensees. It is not clear to the author whether this could compromise the objective setting of water charges.
- Although it was expected that the Bill would include reform of water abstraction, the provisions are limited to removing water companies' statutory right to compensation for losses resulting from their licence to

abstract water. This is in connection with concerns about over-abstraction from early licences granted at a time when there was less concern over low river flows. The EA would be able to make changes to existing licences without paying compensation.

So what are the implications of these major changes? To the author (of this paper) who believes in the power of competition to stimulate innovation and improvements, there are a number of questions and issues:

- The first one is whether it will result in a substantial increase in competition. The experience in Scotland, which introduced competition for non-domestic customers earlier, is that only 5% of commercial organisations have switched to an alternative supplier. As the major cost is in the monopoly distribution function there may be insufficient margins in the other supply costs, particularly as comparative competition has delivered significant efficiency improvements, to make it worthwhile for either new suppliers or for the customers. It may cause incumbents to reduce prices to retain customers, which would be good for those customers but wouldn't reduce the cost of the total service and could threaten sustainability?
- Another question on perceived benefits relates to the additional costs associated with unbundling. There are interface costs particularly those associated with multiple providers of water to the distribution systems. As described earlier in drinking water quality it is a criminal offence to supply "water unfit for human consumption". If a water company is responsible for all aspects from source to tap it is clear who would be responsible. With multiple suppliers, all involved are likely to wish to monitor water quality into the system in case of a major incident. This increases sampling and analysis costs. There will also be costs associated with the interface supply contracts. A report (Investigation into Evidence for Economies of Scale in the Water and Sewerage Industry in England and Wales, 2004) for Ofwat in 2004 had indicated that unbundled interface costs would exceed competition benefits, but the extent to which additional costs offset the perceived competition benefits will only become apparent with experience.
- It is stated in the parliamentary briefing papers that competition has been stifled by the incumbents charging too much for access to the distribution systems. Access charges will be set by Ofwat. With the objective of increasing competition there may be the risk that these charges are set too low with the consequence of too little investment in the underground water and sewerage systems. It is important that access charges fully take into account the capital investment requirements so that refurbishment programmes are maintained both to reduce leakage and for sustainability. Associated with this concern is the lack of Reporters, who had the knowledge to raise issues of sustainability, and whether Ofwat has the water knowledge within its staff to ensure this vital infrastructure investment.

- There are concerns expressed in parliamentary debates on the Bill about incumbent water companies no longer having to be the suppliers of last resort (should newcomers fail to supply). At least one existing water company is indicating that it may wish to opt out of the retail market. What will these changes mean to security of supplies?

3.11 RELEVANCE TO SPAIN?

Without detailed knowledge of the water and wastewater industry in Spain it is not possible to be definitive on the relevance of experience elsewhere. It is possible to discuss those elements which appear to have achieved success in a number of different situations.

Privatisation in England (the change in the status of Welsh Water has already been mentioned) occurred during a period of privatisation of many other public services. Scotland chose to remain public through the formation of a publicly-owned company (a so-called corporatised public) with similar regulation. Would this have been an option for England through creating a commercial operating culture within the public sector? Some other countries looked at the "English model" and started down the privatisation path. For example in the State of Victoria in Australia, small utilities were integrated to form operations of a viable size, and in Melbourne the operation was split to form one wholesale and three retail companies. These were corporatised public companies but with the intention to privatise later. The reforms were highly successful and, as a result, it was considered not to be worthwhile taking a controversial move to privatisation. Chile followed the English model to a certain extent, and now has a mixture of private equity (as in England), concessions (as is common in France) and public, all within the same regulatory system. The regulatory system in Chile was established through the Tariff Law of 1993. This began the reform programme, with one of the major objectives being a transition to full-cost recovery. Importantly it has been a priority in Chile to make provisions for the poor (Rouse, 2013, pp. 209–213). The common theme of the examples of England, Wales, Scotland, Victoria and Chile is effective regulation, not form of ownership. In the cases of England and Chile the conditions required for successful privatisation, in particular water utilities being a viable size, full-cost recovery and strong regulation, were in place ahead of the change date.

The situation in Spain, where responsibility rests with the municipalities, is different from that in England, where water and wastewater service licences are issued by the national government. The important principles of success, which have been present in England (and indeed the whole of the UK) which might apply universally are:

- *Integration of water in national government.* In England all aspects come under one government department, Defra (Department of Environment, Food and Rural Affairs). This integration is extended to regulation with all regulatory agencies coming under Defra (see Figure 3.2).

- *In the author's view, vertical integration of water and wastewater since 1974 allowing optimisation in the supply chain.* Clearly this view is not shared by those advocating unbundling to allow greater direct competition.
- *Regulatory processes which are transparent with documents being publicly available.* This gives the public confidence that development plans and investment programmes are based on need and free from corruption.
- *A periodic planning process in which all parties are involved in an integrated way;* i.e. government (requirements in policy terms), economic regulator (planning coordination and tariff setting), quality regulators (interpreting quality policies into required deliverables and quality improvement programme monitoring), water utilities (business plans and delivery) and consumer bodies (consumer representation). This rigorous planning system ensures that agreed plans are financeable, affordable and deliverable.
- *Good information on asset condition and development costs.* This is vital for success as major investment is required to refurbish existing (and often neglected) infrastructure. Programme and contract failures occur (Rouse, 2013) when the uncertainties associated with lack of or poor information are not taken into account in planning.
- *Full-cost recovery which covers operations, maintenance and refurbishment capital programmes.* This has required political commitment to increases in charges determined by the objective processes. This was not an issue for privatisation as there had been full-cost recovery from charges since 1974. In low cost-recovery situations, in which there has been reliance on grants and subsidies, the transition to full-cost recovery requires a designed period of special transition, or special financing provisions. In Malaysia, there is a government owned organisation called "Water Asset Co" which is able to obtain long-term low-interest loans in the international money market. This organisation finances infrastructure (Rouse, 2014) investments which have been approved by the water regulator in a periodic planning process. Water Asset Co leases the new infrastructure to a water company, typically over 40 years. The leasing charge, approved by the regulator, pays for the asset and ownership is transferred to the water company at the end of the lease term.
- *Efficiency improvements based on regulatory-driven comparative performance.* This raises the question as to whether direct competition, as discussed earlier under Water Bill 2014, will be able to make significant additional advances.

The regulatory process in England and Wales is demanding on the utilities, which require the management skills to respond. Utilities should be of a scale sufficient to afford and attract good management. Although there is a balance between "the carrot and the stick", the approach in the UK is somewhat confrontational. Ofwat is not seen as having any capacity-building role. Where there is a mix between large private and smaller municipal operations, experience has shown that a regulator should be tough when required, but equally have water and wastewater knowledge and experience to provide guidance to the smaller operators. The Portuguese regulator, ERSAR, is an excellent example of such an approach.

The history of water and wastewater throughout the UK shows progressive successful developments over time, irrespective of whether operations were, or are, public or private. Type of ownership or operations is not the issue; the important requirement is to have the fundamentals of good governance (Rouse, 2013) in place.

3.12 REFERENCES

Environmental permitting for discharges to surface water and groundwater http://www.environment-agency.gov.uk/business/topics/water/32038.aspx

Independent Review of Competition and Innovation in Water Markets: Final report: Department for Environment, Food and Rural Affairs PB13690. Crown copyright 2009.

Investigation into evidence for economies of scale in the water and sewerage industry in England and Wales. Report commissioned by Ofwat from Stone & Webster Consultants January 2004 http://www.ofwat.gov.uk/pricereview/pr04/rpt_com_econofscale.pdf

Investing in Scotland's Water Industry: Improvements Delivered in 2006–10. Scottish Government. www.scotland.gov.uk/Resource/Doc/917/0112271.pdf

Ofwat piles pressure on water companies. Financial Times, www.ft.com/cms/s/0/48e00508-8528-11e3-a793-00144feab7de.html (accessed 26 January 2014)

Ofwat stems flow of returns to water investors. Daily Telegraph, http://www.telegraph.co.uk/finance/newsbysector/utilities/10600260/Ofwat-stems-flow-of-returns-to-water-investors.html (accessed 27 January 2014)

Okun D. (1977). Regionalisation of Water Management, A Revolution in England and Wales. Applied Science Publishers, UK.

Relative efficiency assessment 2008–9 – supporting information. PR09/39 Ofwat Dec 2009. http://ofwat.gov.uk/publications/pricereviewletters/ltr_pr0939_appendix2.pdf

Review of Ofwat and consumer representation in the water sector. Department for Environment, Food and Rural Affairs PB13587. Crown copyright 2011

Rouse M. J. (2013). Institutional Governance and Regulation of Water Services: The Essential Elements, 2nd edn. IWA Publishing, London.

Rouse M. J. (2014). The worldwide urban water and wastewater infrastructure challenge. International Journal of Water Resources Development, doi: 10.1080/07900627.2014.882203

Select Committee on Environmental Audit Minutes of Evidence Memorandum from the Office of Water Services (Ofwat) http://www.publications.parliament.uk/pa/cm200001/cmselect/cmenvaud/290/1022802.htm

Service and Delivery: Performance of the Water Companies in England and Wales 2009–10 Report. Ofwat, UK. www.ofwat.gov.uk/regulating/reporting/rpt_los_2009-10supinfo.pdf

Setting price controls for 2015–2020 – risk and reward guidance, January 2014. www.ofwat.gov.uk

Taken for granted. Report of the Working Party on Sewage Disposal. HMSO 1974 ISBN: 0117502200.

The Development of the Water Industry in England and Wales. Defra and Ofwat 2006 Crown Copyright.

The Independent Review of Charging for Household Water and Sewerage Services. Published by the Department for Environment, Food and Rural Affairs. PB13336 © Crown Copyright 2009.

Water Bill 2013–14. UK Parliament. www.parliament.uk/briefing-papers/lln-2014-002/water-bill

Water Industry Act 1991 The Stationery Office ISBN: 0-10-545691-8.

Water Management. House of Lords Science and Technology Committee Report. The Stationery Office. June 2006. ISBN: 100104008717.

Water prices in England and Wales from April 2007. Water UK Press Release 26 February 2007 www.water.org.uk

Chapter 4

Water regulation in Australia

Andrew Speers

Chair, Centre for Comparative Water Policy and Laws University of South Australia

4.1 INTRODUCTION

This paper is concerned with the direction of water regulation in Australia, past and present. It is intended to provide an understanding of the reasons Australia embarked on a major process of water reform in the 1980s and the characteristics of that reform. Broadly speaking, the goals of this reform process were: to improve the efficiency with which water services were delivered; to reduce dependence on debt as a means of financing capital assets; to improve the sustainability of water services; and to ensure that prices charged for water services reflected the costs of providing those services. Each of these goals is explained in this paper.

The reform process created new institutions with new responsibilities which needed to be regulated in new ways. Accordingly, the regulatory framework was revised to ensure that the responsibilities of all institutions involved in the management of urban water supplies were clarified and appropriately assigned, and that the market power of newly created water supply companies was efficiently managed. Other aspects of regulation ensured that consumer rights were protected, and that the environment on which water supply companies so critically depend was not compromised. The new regulatory framework is described in this paper beginning with an explanation of the regulatory principles agreed to by all Australian governments, and the various models employed by particular governments to implement those principles.

Discussion of price regulation is a particular feature of this paper. There are two reasons for this. Firstly, most water supply companies are monopolies within the areas in which they operate. Australian governments agree that where there is a lack of competition, the prices charged by water suppliers should be subject to regulatory controls so that market power is not abused. The paper, therefore, includes discussion of the principles of price regulation and the regulatory framework for price control. Secondly, one of the ongoing goals of the water reform process in Australia is to promote competition. A more competitive environment will tend to put downward pressure on prices without government intervention and thus the regulatory environment for prices continues to evolve. Some of the approaches that are now being discussed in Australia to respond to the existence of greater competition are therefore also explored.

Australia is a federation divided politically into six states and two territories, each of which is self-governing, and there is a federal (national) government. While each jurisdiction has agreed to various principles for water reform, the way each government has responded to these principles has varied. Thus, it is possible to make some observations about which approaches have been the most successful, and which have been abandoned or modified in the light of experience. This paper includes some observations about approaches that have worked and those that have produced less effective outcomes in the hope that these may prove valuable to Spanish authorities as they embark on their own process of regulatory reform.

4.2 WATER RESOURCE CHARACTERISTICS

Anyone who has ever visited Australia, whether on water-related business or not, will sooner or later be told that Australia is the planet's driest inhabited continent. This is true, but the statistic is largely meaningless. Parts of Australia in fact have very high annual rainfalls, up to or exceeding 4000 mm per annum, although desert areas, which comprise a very large proportion of the country commonly receive less than 150 mm per annum. Figure 4.1 shows the pattern of rainfall across the nation, illustrating the wide variability. One of the valuable interpretations that can be taken from Figure 4.1 is that 60% of water runoff in Australia occurs in the tropics, with only 6% occurring in the Murray-Darling basin (in the south-east of the continent) where 50% of Australia's water use occurs.

In addition to understanding the physical pattern of rainfall and runoff one must also consider the temporal variability of rainfall. In the tropics (across the north) most rainfall is concentrated typically in a three-month period or shorter. Across much of the rest of the country, rainfall varies considerably from year to year with extended periods of drought not uncommon. Figure 4.2 shows the variability over a 97-year period of rainfall in the city of Perth, Western Australia. Most major urban areas experience similar conditions. The securing of permanent water supplies in the face of such variability has always been the most significant urban water supply challenge Australia has faced.

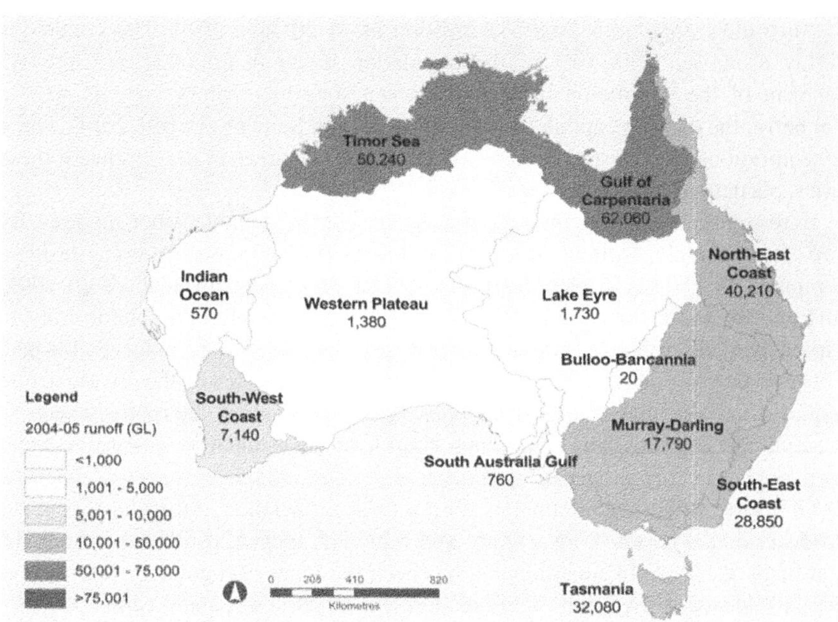

Figure 4.1 Australian runoff volumes 2004–05 from each drainage division (National Water Commission, 2005a).

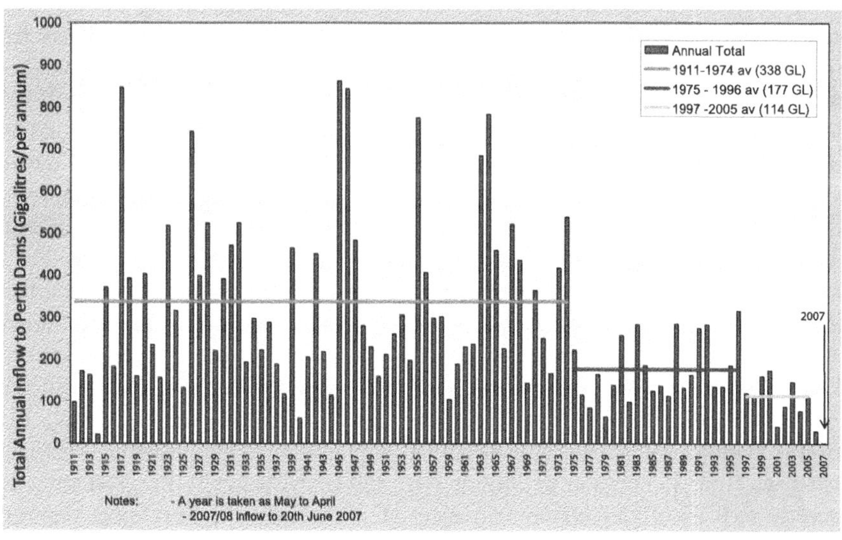

Figure 4.2 Trend in inflows to surface water storages in Perth, Western Australia. Source: Water Corporation.

Australia's population was 23.1 million as of 30 June 2013. The country is highly urbanised with 89% of the population living in urban areas. Sixty-six per cent of the population lives in the seven state or territory capitals, and in Canberra, the nation's capital. (Australian Bureau of Statistics [ABS], 2014). These concentrations place large pressures on the water resources in the regions of these cities, particularly in dry periods.

In response, cities have invested heavily in securing reliable water supplies. By 2002, for example, Sydney stored 932 kilolitres (kL) of water for every resident compared to 250 kL in New York and 182 kL in London (Tisdell *et al.* 2002) sufficient to see it through a severe and prolonged drought. The opportunity to exploit available water resources further as populations grow is, however, limited. Those resources that can be accessed are increasingly remote and the infrastructure required to store and transport supplies is costly – as are alternatives such as desalination and recycling – although significant investment in innovative water facilities has occurred over the past decade.

Of course urban water supplies do not exist in isolation. Australia is a major producer and exporter of agriculture and minerals. Indeed, rural water use is the dominant water use in Australia and has been the focus of regulatory reform over the past decade. Urban users share resources with rural users and competition for the resource has emerged and has been the subject of political controversy. This has affected the way in which water suppliers in urban areas have managed and expanded their water supplies to meet the needs of growing populations.

Australians in urban areas are almost universally provided with reticulated water services, with the coverage for sewerage services being about 97%. Regulatory reform (including price) coupled with very severe drought has led to more conservative behaviour by consumers. This trend continues an earlier pattern than saw a reduction in average household water consumption of 21.4% between 2003–2007. Total consumptive use of water in the Australian economy in 2004–05 was 18.8 million gigalitres (GL) which is 6.4% of the resource. Urban water consumption (household, water supply and manufacturing) comprised 25.5%, of this total. The largest water user was the agriculture sector (65% of total consumptive water use) (NWC, 2005a, 2005b).

In response to drought, competition for water resources, and a growing urban population, major capital city water suppliers have sought to diversify water supplies. Traditionally, urban suppliers have relied on surface or ground waters, depending on the resources available. Over the past 20 years, however, significant investment has been made in water recycling and in desalination. For example, the installed production capacity of desalination plants in capital cities in Australia increased from 45 GL/year in 2006 to over 450 GL/year by 2013 representing an increase to 9.4% of total supply (Hoang *et al.* 2009). As with other water sources, there is considerable variation in production capacity between cities. The city of Perth, Western Australia, is now able to meet as much as 50% of its supply need from desalination if necessary.

Similarly, the installed capacity of reclaimed (recycled) water has increased substantially over the past 10–15 years. While use of this water source tends to vary significantly in response to climatic conditions (for example, because rainfall makes irrigation unnecessary) relatively recent data suggests that the proportion of recycled water provided by urban water suppliers may be as high as 19.4% of total supply (Radcliffe, 2006). Recycled water in Australia is currently only used for non-potable purposes, although indirect potable reuse often occurs.

4.3 CATALYSTS FOR REFORM

The objective of water management policy from the colonial period (beginning with European settlement in 1788) to the 1980s was to exploit available water resources to provide for agriculture, urban development and industrial production. Little thought was given to global limit of the accessible water resources. Consequently, the tendency was to access more remote water sources, once local sources had been fully exploited. As the resources are limited by their annual yield (replenishment rate), this is an approach which cannot continue endlessly. By the late 1970s-early 1980s economically exploitable water resources were approaching exhaustion in most major urban centres. Furthermore, investment in infrastructure renewal had been low as capital had been directed to securing water supplies, rather than replacement of existing infrastructure. The result was a rising marginal cost of water supply and competition between users for access to water (Tisdell *et al.* 2002).

The exploitation of water resources was also resulting in environmental degradation of the resources themselves, the riparian ecosystems dependent on them, and downstream environments to which wastewater was discharged. Eutrophication of water ways, increased intermittency of flows, contamination by industrial pollutants and the integrity of ecosystems were all issues of significant concern. Thus, for the water industry, the focus moved from resource development to water management. In other words, water and related infrastructure had to be administered to achieve multiple objectives including efficient use of capital, equitable sharing of the resource and ecological sustainability.

The water industry is, of course, only a subset of the economy and wider influences also affected it. While Australia had experienced favourable economic conditions for the three decades following World War II, these deteriorated in the early 1970s with the emergence of high unemployment and inflation. Government policy to this point had been concerned with macro-economic conditions, and micro-economic reform – particularly that directed to improving the efficiency of Australian industry – had withered. In response, in the early 1980s the Australian federal government began a major programme of micro-economic reform which included the abolition of exchange controls, floating of the Australian dollar and reduction in tariffs. This exposed Australian industry to increased international competition. One of the outcomes of this exposure was a new focus on the costs of regulation and the cost of inputs – such as water services – provided by government business enterprises (GBEs) which included the nation's major urban water suppliers. Thus, micro-economic reform was extended to exposing GBEs to market forces and in 1993 a report by the National Competition Policy Review Committee (NCPRC) recommended, among other things, that GBEs that controlled an "essential facility" with natural monopoly characteristics be reformed to ensure that their market power was not abused (for example, by setting prices above the long-run marginal cost of production, or by regulating against competition or by discriminating among customers on the basis of price). Reform of GBEs was a component of what became known as National Competition Policy which emphasised open markets and competitive neutrality (between government-owned businesses and private industry) including transferring of activities to the private sector by "contracting out" opportunities to construct and/or operate facilities and services.

4.4 COUNCIL OF AUSTRALIAN GOVERNMENTS WATER REFORM FRAMEWORK 1994

The NCPRC report became commonly known as the "Hilmer Report" after its Chair, Professor Fred Hilmer. In 1994, in response to the Hilmer Report, the Council of Australian Governments (COAG), which includes the Australian federal government, all state and territory governments, and a representative of local government, unanimously endorsed the COAG Water Reform Framework. The principles of the Framework relevant to urban water reform included (Council of Australian Governments, 1994):

- The creation of water supply utilities that are commercially focused. This was to be done by either establishing these utilities as government trading enterprises (that is 'corporatising' them) or fully or partially privatising them;
- The implementation of full-cost recovery such that the prices charged to consumers reflect the cost of providing the service, including a real rate of return on the written down replacement cost of assets (profit);
- The adoption of pricing regimes which reflect water consumed (consumption-based pricing);
- The removal of cross-subsidies between user groups (such as between industrial and residential users);
- Where subsidies are required for social-benefit purposes (for example to assist low-income users) the subsidies were to be transparent and to be reimbursed by governments to water suppliers;
- Institutional separation of the roles of water resource management, standard setting and regulatory enforcement and service provision; and
- Benchmarking water supply utilities against each other to encourage "competition by comparison" and fostering a move toward world's best practice.

The COAG Water Reform Framework created a model for all governments to follow, although each State and Territory government was responsible for deciding how best to achieve each of the principles. In practice, some governments' achievements were more substantial than others. No government elected to privatise its water utilities, but each sought to introduce competitive pressures through various means, mainly by establishing the utilities as government-owned businesses subject to the same regulations as any private company. The COAG Water Reform Framework included numerous measures directed at improving the performance of rural water supply schemes (primarily irrigation schemes) and at clarifying ownership and entitlement arrangements for non-urban water. It is in the management of urban water, however, that the Framework had its most significant impact.

That much was achieved as a result of the Framework's implementation is testament to the commitment of governments to reform during this period. However, additional stimulus was provided by the Federal government which offered substantial financial incentives (payments) to the States and Territories upon achievement of various milestones associated with the reforms. For example, the 1994 Framework committed governments to achieving cost-reflective pricing by 1998. Once this milestone was achieved, the Federal government released promised funding.

It is now 20 years since the Framework was promulgated. It has resulted in significant efficiency gains throughout the water industry; it has improved investment in (and subsequently the condition of) water infrastructure; and it has introduced innovation and competition within the industry. Its principles remain the basis of urban water policy in Australia today.

4.5 THE NATIONAL WATER INITIATIVE

Whereas the Water Reform Framework laid the foundations for water reform in Australia, the National Water Initiative (NWI) expanded the scope of the Framework to cover more fully rural water, and to bring consistency in water management – urban and rural – across the country. Most of the NWI is concerned with rural water management, which is beyond the scope of this paper. Nevertheless, as stated earlier, water used in urban areas does not exist in isolation from other environments and in most cases is shared with rural water in a number of ways. Thus, elements of the NWI that are listed below include some items that may appear to have a more rural focus, but in fact are just as relevant to urban water as they are to rural. Furthermore, because the rural water issues covered in the NWI are so extensive and fundamental a brief comment is also included about the nature of rural water reform in Australia.

The NWI was signed in 2004 by most Australian governments (the government of Western Australia signed in 2005). Water, however, is no respecter of political boundaries. While urban water might be managed on a state or territory basis, rural and riverine systems require cooperation between jurisdictions. The National

Water Initiative is Australia's blueprint for national cooperation. The broad objectives of the NWI agreed to by the governments are to create a "nationally-compatible, market, regulatory and planning based system of managing surface and groundwater resources for rural and urban use that optimises economic, social and environmental outcomes" (Council of Australian Governments, 2004). The National Water Commission (NWC), established to oversee implementation of the National Water Initiative, states that the NWI sets out the protocols through which Australia manages, measures, plans for, prices, and trades water (see www.nwc.gov.au/nwi).

A summary of the NWI elements concerned with urban water management includes the following (Council of Australian Governments, 2004):

- The preparation of water plans for each catchment, specifying the amount of water that can be withdrawn, taking into account social and environmental objectives;
- Clarification of water entitlements such that ownership is clear and risks are appropriately allocated;
- Creation of a market for trading of water between entitlement holders, including between urban and rural areas;
- Development of a system of water accounts to ensure that adequate measurement, monitoring and reporting systems are in place in all jurisdictions to support public and investor confidence in the amount of water being traded, extracted for consumptive use, and recovered and managed;
- Work toward improving water pricing regimes to ensure there is full cost recovery of urban *and* rural water systems, including the costs of planning, and incorporating, in urban settings, a price for recycled water and stormwater use by consumers. This ensures that a price is set for water provided by supply utilities regardless of its source;
- Internalisation of environmental externalities in price, or through regulation depending on feasibility; and
- Promotion of water efficient urban design and implementation of demand management practices in urban areas to promote efficient use of the resource and taking into account all available sources of water (potable water, recycled water, stormwater, desalinated water, wastewater).

As will be noted, there is considerably more detail in the NWI than the Framework, but this reflects experience and a growing awareness that rural water management had not advanced as rapidly as urban water management. While less attention was paid to urban water in the NWI, the fact that water plans, water entitlement regimes, and water markets were all required under the NWI affected the management of urban water. For example, it became possible for urban areas to trade purchase water entitlements from rural areas.

The NWI did not include payments for achievement of milestones as the Framework had, but a National Water Commission was established to oversee implementation of the NWI. The NWC does not have regulatory authority, but is required to report biennially (now triennially) on progress against the NWI goals, enabling it to use moral force to encourage compliance with the NWI goals.

4.5.1 The regulatory environment for water

The COAG Water Reform Framework and the National Water Initiative that builds upon it guide the development of the regulatory regime for urban (and rural) water in Australia. They have been set out in this paper to provide insight into their principles and elements relevant to urban water against which further description can be offered. The remainder of this paper, then, explores the regulatory objectives, the approaches taken by governments to achieving them, achievements, pitfalls and areas for improvement. Before beginning that task, however, it needs to be made clear that discussion about regulation is impossible without parallel discussion of the way in which price has been used as a mechanism to rationalise water consumption and investment in water utilities. Thus, the following also includes considerable analysis of matters to do with price reform.

4.5.2 Institutional arrangements

Historically, urban water utilities had been created and operated by state, territory or local governments as government departments. That is, they were subject to the direction of state or territory Ministers, or local governments. Not uncommonly, capital was transferred freely between the utility and the government to which it was responsible and investment decisions were often subject to political exigencies. As urban development is dependent upon water and sewerage services, the utilities themselves often became quite powerful determining patterns of urban development according to their own needs, rather than rational cost-benefit analysis. As a result, while almost universal coverage was achieved for water and sewerage services in urban areas, investments were often poorly made (resulting in greater costs than benefits), debt levels were high, investment in infrastructure renewal was low and, generally, significant inefficiencies existed.

Additionally, there was significant confusion and overlap between agencies responsible for the supply of water, health and safety standards and pollution control. For example, until the late 1980s the Chief Executive of the, then, Sydney Water Board sat on the Board of the State Pollution Control Commission, an obvious conflict of interest. Furthermore, government authorities such as these were not subject to the same regulatory regimes that applied to private companies, significantly reducing the costs they faced and placing them at a competitive advantage over private firms which might otherwise have tried to compete in the water market. Finally, government-controlled water authorities were instruments of government policy which distorted the objectives they were established to achieve. It was standard practice, for example, for utilities to charge the commercial-industrial sector significantly more for water services than they did the residential sector. The idealist might suggest that this kept costs to the less-advantaged as low as possible; the cynic might observe that companies don't vote.

The primary reform goal then was to establish utilities as separate businesses enterprises accountable for their own performance and responsible to independent boards in the manner of a private company. The details of the approach taken varied from jurisdiction to jurisdiction, but the essential elements included:

- *Establishment of utilities as corporations*, rather than government departments. This, and other legislative changes, ensured that utilities were subject to all laws, with no greater protections, exemptions or obligations than any private company. Importantly, this meant that the utilities were subject to tax on profits and were obliged to ensure that they were solvent, obtaining funds required to fund their operations through their own capital raising, or from charges levied on consumers. Any requests by government to achieve particular goals, or to reduce costs to under-privileged consumers, were accompanied by payments to fund those objectives, which were explicitly identified in the utilities' accounts.
- *Creation of "shareholders"* with expectations of return on investment. In practice, the shareholder was the government, represented by relevant Ministers. For example, in the state of New South Wales the Treasurer, the Minister for Finance and the Minister for Water were the three shareholders in the company. In other jurisdictions, the entity which initially owned the water supply system, for example, a local government, became the shareholding entity. As is the expectation of the shareholders of any private company, the utilities' shareholders expect that the company will be profitable and will deliver a dividend annually representing that proportion of the profit the utility's board decides is available (that is, not required for reinvestment or to retire debt) for distribution to shareholders. In theory it is the board that makes this decision, in practice there are usually negotiations between the shareholders and the board which, at times, has produced non-commercial decisions (discussed further, below).
- *Payment of tax equivalents*. Government owned organisations are generally not subject to federal income tax. If they were not to pay such tax they would be at a competitive advantage over private companies. Consequently, utilities now pay an amount equivalent to that which they would have paid to the federal government to their state/territory government owners. Thus, utilities' payments to their governments include a dividend, and a payment under the "tax equivalence regime" (TER). For large companies, such as Sydney Water which provides water services to some 5 million people, the value of the dividend and the TER can be in the A$ Billions per year (A$ 1 = € 0.67 approximately).
- *Establishment of objectives*. Usually incorporated in an Operating Licence or similar instruments intended to define the role of the organisation. For

example, in a competitive market, an organisation may be able to increase its market share by going beyond regulatory requirements and promoting itself as a "good corporate citizen". Where there is no competition, there is no opportunity to gain market share and thus no incentive to go beyond the minimum. Thus, Operating Licences set out objectives the organisation is required by governments to meet. Usually these are renewed periodically and may be audited publicly. In theory, where a utility has failed to meet its objectives its Operating Licence could be ceded to a private organisation which would take over operation of the supply system. While beneficial, Operating Licences have proved imperfect in some instances as some governments have attempted to load these instruments with too many subordinate requirements, reducing utilities' focus on their core objectives of delivering high quality water and sewerage services. This problem too is discussed further below.

- *Structural separation of regulatory and operational roles.* Prior to the implementation of the Framework water supply, environmental and health regulation functions tended to be intertwined. As indicated earlier, for example, there was not always clear separation between pollution control responsibilities and water supply, and water suppliers often had a role in specifying the health standards that would be achieved. Furthermore, the prices set by utilities was often set by the utility itself, or in consultation with government, leading potentially either to under-pricing of services to meet governments' political or social security objectives, or over-pricing in line with the common practice of monopolies. Often, also, utility directors were drawn from the ranks of government owners, obscuring the role of the utility *vis-à-vis* government. Institutional reform ensured the separate of these entities. Thus, in New South Wales, for example, Sydney Water became responsible for water, sewerage and some drainage services in the greater Sydney area, the Environment Protection Authority became responsible for water pollution control, the National Health and Medical Research Council and the state Department of Health for water related health standards, and the Independent Pricing and Regulatory Tribunal for price setting[1].

- *Separation of water supply from water resource management functions.* While not specifically a requirement of either the Framework or the NWI, several jurisdictions have separated natural resource management functions from water supply functions. For example, in the state of Victoria, Melbourne Water has been established as the bulk water supplier to the three government-owned water supply companies that service different parts of Melbourne.

---

[1]Note that this is a simplified view. There are other agencies responsible for aspects of regulation depending on the circumstance (for example, local government approval is often required for construction of works).

Similar models have been established in most other states and territories. The idea behind this supplementary reform is that natural resource protection objectives may conflict with the objectives of water supply (i.e. there may be a conflict between protection of rivers and exploitation of resources to generate sales) and should therefore be separated.

Unsurprisingly, details of the models used varied from jurisdiction to jurisdiction, although the intent of reforms was to achieve the requirements of the Framework and later the NWI. Each jurisdiction also initially adopted quite divergent approaches to achievement of requirements, although over time there has generally been coalescence around a model including versions of the institutions described for New South Wales, above. Figure 4.3, below, shows the current institutional circumstances in each state and territory.

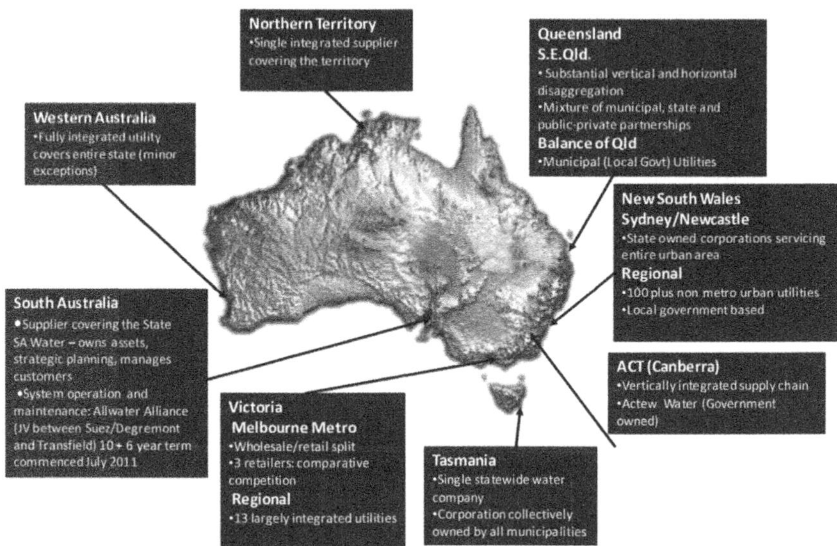

Figure 4.3 Current institutional arrangements for urban water management in Australia.

4.5.3 Water pricing

It is difficult to overestimate the importance of price reform in Australia in the economically viable operation of utilities, or the rationalisation of water use in urban areas. It is, of course, a basic principle of economics that if a good or service is over- or under-priced, it will be over- or under-consumed. Both circumstances represent inefficiencies. Before the implementation of pricing reform, both circumstances existed in Australia. For example, under-pricing of water had led to unsustainable increases in the volume of water demanded by residential consumers which, in turn, tended to lead to under-investment in infrastructure renewal, or the assumption by utilities of high levels of debt (or the injection of capital by governments, depending on the circumstances). Conversely, unreasonably high charges levied on commercial/industrial premises which did not reflect the volume of water consumed and over which the premises had no control was a cost to companies that prevented them from making more productive investments and reduced their competitiveness. Furthermore, even for residential customers, the price paid was commonly based on property values. It was therefore a tax, which did not reflect the volume of water consumed. Thus, there was no price signal to customers to rationalise water consumption. In summary pricing that was not "cost-reflective" was leading to unsustainable levels of demand for water, under-investment in infrastructure renewal and maintenance, high levels of debt and a cost to industry that reduced competitiveness.

It follows that if the intention of the 1994 Framework was, in part, to have utilities operate as businesses, the prices they charged had to cover the long-run marginal cost of production. As a supplement to the Framework, COAG established pricing principles to guide the way in which prices were calculated, whether by an independent economic regulator or by government where such a regulator did not exist. The principles stated that a viable water business should recover the "operational, maintenance administrative costs, [environment and natural resource] externalities, taxes … or TERs, the interest cost on debt, dividends (if any) and make provision for future assets refurbishment/replacement". Further, "dividends should be set at a level that reflects commercial reality and stimulates a competitive market outcome" [2]. Other principles included instructions on the prices that could be charged to avoid monopoly behaviour by utilities and instructions on the way economic regulators should incorporate recognition of efficient resource pricing and business costs (Council of Australian Governments, 2010, p.18). Summarising the intent of water price reform, the National Water Commission said:

> "Efficient pricing or charging for water-related services underpins investment and provides signals for the efficient use of water services. Getting the price signals right by ensuring that they fully reflect the efficient costs of providing the services is a key element in encouraging innovation and efficient water use" (National Water Commission, 2011, p.11).

Pricing reform, particularly during the early period, was not without its critics, some of whom argued that water was a public good whereas the approaches taken priced it as though it were a private good. Much of that criticism has abated over

---

[2]In these principles, the deprival value methodology is used to calculate the value of an asset (that is the value of the loss that would be suffered if the owner were deprived of the asset). Also, the annuity approach is used to determine the medium- to long-term cash requirements for asset replacement/ refurbishment where it is desired that the service delivery capacity be maintained (Council of Australian Governments, 2010, p.18).

time. More relevant is the observation that not all utilities were equally well positioned to move to full cost-reflective prices rapidly. Indeed, the process is ongoing to this day. While the major capital city utilities moved more rapidly to cost-reflective pricing, reform of other, mainly non-capital city urban suppliers has occurred more slowly. Additionally, over time there has been refinement of the items to be included within the prices charged. So, for example, pricing principles have now been promulgated that guide utilities (or their regulators) in incorporating the cost of water service planning and management activities in the price of water and guide them in the setting of prices for use of stormwater and recycled water (Department of Environment, Water, Heritage and the Arts, 2010). Furthermore, the extent to which water suppliers are in a position to provide a dividend or pay a TER varies, particularly between rural and urban areas, and the basis of calculation of asset valuation and the cost of capital has been refined over time. Presently, while all utilities are expected to work toward "upper bound pricing" which is the limit of cost recovery beyond which price moves into the territory of monopoly rents (as opposed to "lower bound pricing" which is the minimum level of cost recovery necessary to ensure the viability of water businesses).

In summary, the intent of water price reform was to:

- Ensure that prices are set in a way that leads to the economically efficient use of water and water service provision;
- Ensure the financial viability of water businesses and the effective regulation of monopolies; and
- Maintain equity in a transparent manner by separately identifying any subsidies paid to lower socio-economic status groups to avoid distorting efficient pricing outcomes.

4.5.3.1 Economic regulation

As has been discussed, the lack of direct competition in the market for supply of water to urban areas increases the potential for monopoly pricing to occur. Benchmarking and efficiency goals included in Operating Licences provide a partial substitute for a lack of market discipline. A further, important, element is the existence of economic regulators, the basic role of which is to set the maximum prices that can be charged for water and related services (sewerage, stormwater, recycled water etc.). Economic regulators are established by, but theoretically independent from, government (that is, not subject to direction of a Minister in reaching their determinations).

Despite decades of water reform not all states or territories have established such entities. Nevertheless, three sophisticated examples are the Essential Services Commission in Victoria (www.esc.vic.gov.au/water), the Independent Pricing and Regulatory Tribunal in New South Wales (www.ipart.nsw.gov.au) and the Economic Regulation Authority in Western Australia (www.erawa.com.au). The methodologies and scope of review of these bodies varies from jurisdiction to jurisdiction (for example they may review the performance of utilities or set prices for other services such as drainage). Each, however, responds to the basic pricing principles for water supply set out in the 1994 Framework, and the National Water Initiative.

4.5.4 Benchmarking

The 1994 Framework proposed that urban water utilities should be compared, one to another, to simulate competition and foster world's best practice in respect to water services. The National Water Initiative extended this idea, and made it a formal component of the Initiative. Consequently, all major urban water utilities participate in regular benchmarking activities against a range of criteria, including water resource management, pricing, finance, customer service, environment and health, and bulk water supply. There have also been, at times, some occasional indicators, or indicators that are only measured periodically, such as the volume of recycled water supplied. Originally these reports were published by the Water Services Association of Australia (WSAA) which represents the major water suppliers nationwide. Responsibility is, however, now shared with the National Water Commission. The 2012–13 *National Performance Report* was the eighth report in the series (National Water Commission, 2013). Reports are compiled approximately every two years. Currently there are 81 participating utilities including all of the major capital city and non-capital city utilities, and some smaller town suppliers. These utilities provide supply to 18.7 million Australians (81% of the population).

The *National Performance Report* (which is complemented by a similar report covering rural water service providers, such as irrigation schemes) allows comparison between systems on a common basis. While there remains very little direct competition in the market for urban water services, these benchmarking processes enable a level of "competition by comparison". While geographic and other conditions limit the validity of comparisons in all circumstances, utilities are highly aware of the relative performance of comparable organisations. Such comparison has helped to drive performance, such that Australian utilities can now boast, for example, 100% compliance with Australian Drinking Water Guidelines and eight can report that they now recycle 90% of the effluent for which they are responsible. The *National Performance Reports* are also valuable in explaining trends within the industry. For example, the average cost per customer has risen substantially over the past several years for most utilities so customers are able to see that there are forces external to the industry – in this case drought and the concomitant cost of securing water supplies from less traditional sources – which have an impact on all suppliers. Individual state or territory based economic regulators, such as the Essential Services Commission in Victoria may also compile reports comparing the performance of utilities (see for example http://www.esc.vic. gov.au/Water/Performance-reports).

The industry itself, through the Water Services Association of Australia, has also participated in benchmarking activities beyond those mandated by the National Water Initiative. For example, in late 2012 WSAA completed the *International Asset Management Performance Improvement Project* to examine the relative performance of asset management across utilities in various countries, as part of the effort to achieve world's best practice. Thirty-seven utilities from Australia, Canada, New Zealand, the Philippines, and the United States participated in this initiative. The project was jointly sponsored by the International Water Association. The 2012 report was a repeat of an initiative compiled in 2008. There was a trend in the intervening period toward improved asset management performance (Water Services Association of Australia, 2012).

While not specifically a benchmarking activity, the National Water Commission also has legislative responsibility for compiling biennial (now triennial) reviews of performance against the National Water Initiative. These reports do not compare utilities to each other, but rather analyse the national picture with respect to achievement of the NWI. A broad picture is provided, as well as state/territory based analyses of progress. The independent nature of the National Water Commission has allowed it to be forthright and to offer criticisms when they are due. Again, the Commission is using moral suasion to encourage governments to commit resources to legislative reform to achieve the commitments they have previously made under the National Water Initiative. The reports, and progress, are well covered by the media, although such coverage could probably not be expected in a country in which water management was not a widely discussed topic in the community. The last Biennial Assessment was completed in 2011, the first Triennial Assessment will be available this year. The Assessments are submitted to the Chair of the Council of Australian Governments (the Prime Minister) rather than any particular government, or Minister.

Various benchmarking activities have been important steps in fostering compliance with the goals of the Framework and the NWI and in improving performance of individual utilities. It is notable that to improve their performance utilities have initiated their own benchmarking activities beyond the requirements of the NWI. That these additional activities are focused on matters that have the potential to affect utilities' bottom-line significantly is testament to their willingness to behave as any private company would in reducing costs and improving efficiency.

4.5.5 Environmental and health regulation

Institutional reform, as explained, has led to separation of operational responsibilities from regulatory responsibilities. The objectives of water utilities are to supply water and sewerage services and in some circumstances, drainage services. Other agencies are responsible for environmental regulation or for the establishment of drinking water quality standards. These approaches eliminate the conflicts of interest that had formerly existed.

Generally speaking, environmental regulation is the domain of state- or territory-based agencies. These organisations commonly set discharge standards for effluent, or matters relevant to the operation of sewerage systems (for example, the frequency or volume of discharge allowed during wet weather). Increasingly, environmental authorities have sought to use economic instruments to drive improved performance. For example, in one case several sewage treatment plants discharging to a river were brought under one licence with the utility free to decide where to invest to achieve a global discharge standard. That is, the utility could decide to upgrade one plant only and leave the others to discharge at a lower standard as long as a global water quality standard was maintained. This provided opportunity for the utility to determine the best investment strategy, reducing inefficiencies.

Environment agencies may also be involved in the setting of trade waste discharge standards, although this varies from circumstance to circumstance. In some situations, utilities assume this responsibility reflecting the fact that they must meet their own downstream discharge standards and that the control of trade waste is a means of achieving those standards. Commonly, economic instruments are used to encourage compliance. Thus, trade waste discharge fees are based on the quantity and strength of the discharge with charges increasing exponentially to encourage moves to lower strength effluents. Some substances are banned outright from discharge by the environmental agency or a utility may refuse to accept them.

4.5.5.1 Water trading

To understand fully the water industry in Australia, the development of a water market must also be understood. While water markets primarily affect rural water allocation, urban water managers are not uninvolved, particularly as the market envisages the potential for trading between rural and urban areas (or vice versa) to occur. It must be noted that in Australia land and water titles are separate. That is, a landowner only has as much right to water on his or her property as he or she requires for "stock and domestic" use[3]. Additional water as may be needed (say, for irrigation) has to be secured through purchase of water entitlements on the open market. These entitlements can be traded more or less freely and the price fluctuates according to demand. The federal government is a participant in this market, buying water entitlements on behalf of the environment, and thereby returning over-exploited catchments to more sustainable conditions. Each catchment area is to be subject to a Water Plan (described below) describing the limits of extraction in any catchment area and regularising entitlements. While in

---

[3]This term refers to water use for a household, a kitchen garden, pets and grazing cattle or sheep (or similar livestock). It does not include water for intensive agriculture (e.g. piggeries) or other commercial uses.

theory the market is free and open, governments have tended to unwisely impose limits on inter-basin trades of water and on trades from rural to urban areas. (See http://www.nationalwatermarket.gov.au/ for more information on Australia's water markets.)

4.5.5.2 Water plans

At a far more fundamental level a principal instrument guiding the sustainable management of water resources is the development of Water Plans as required under the National Water Initiative. Water Plans are legally enforceable instruments that establish the total allowable volume of water that can be extracted from a source (surface or groundwater) for consumptive use, and sets out the arrangements for sharing the water available among competing users. The volume of water that can be extracted will generally vary in the plans in response to conditions. For example, volumes available may be reduced during dry periods in which case the category of entitlement an entitlement holder owns will dictate the amount of water they can use. Except in more extreme droughts, high-security entitlement holders will be able to use the volumes of water allocated to them, where as lower-security entitlement holders will face limits on the volume of water they can use. The environment itself is generally specified as a high-security entitlement holder. Of course, entitlements can be traded on the open market.

The intent of statutorily-based Water Plans is described by the Australian Government's Department of Environment as:

> "… the vehicle for the setting of sustainable environmental, social and economic objectives for the management of water resources. Effective water plans establish the rules to meet environmental objectives and for users to share water resources by providing certainty of access to a share of water over an agreed timeframe" (Environment.gov.au. 2014).

Water Plans cover both rural and urban areas. In urban areas water bodies have often been substantially modified, and urban demands may be the dominant use within any one catchment. Nevertheless, the environment does have a claim to water and the plans deal with environmental water demand in increasingly sophisticated ways (for example specifying variability in environmental releases to match naturally occurring variability in river flow). Furthermore, resources in any catchment need to be shared between urban and non-urban uses and so it remains important that entitlements be specified and secure, including entitlements held by the environment itself. The extent to which there is competition for the resource between urban and rural uses varies across catchments. In some areas in which urban centres exist within largely rural environments there may be significant competition for water. Useful insight to the operation and detail of Water Plans, and the status of Plans across the nation is available on the National Water Commission's website (National Water Commission, 2012).

4.5.5.3 *Water accounting*

A component of the National Water Initiative is the preparation annually of a National Water Account. According to the Australian Bureau of Meteorology the National Water Account:

> "... provides a detailed insight into the management of Australia's water resources at the national and regional scale. It supports the National Water Initiative, disclosing the total water resource, the volume of water available for abstraction, the rights to abstract water, and the actual abstraction of water for economic, social, cultural and environmental benefit across Australia" (Bureau of Meteorology, 2013).

The National Water Account provides standardised information that meets a specified Water Accounting Standard. It is compiled annually and the information within it is publicly available. The National Water Account includes the following information:

- Changes in water inflows, outflows and storages that occurred over the reporting period;
- The water access entitlements that existed during the period;
- The Water Management Plans that applied within the period to govern water access;
- The volume of water allocated for use during the period;
- How much of the total water entitlement was traded during the period;
- The volume of water abstracted for use; and
- The volume of water made available to the environment.

The National Water Account is of day-to-day importance to rural users, particularly irrigators or investors in water entitlements. It is also, however, relevant to urban utilities as it specifies the size of the resource from which they may be entitled to draw, changes in availability through the reporting period and other uses within the catchment.

4.5.6 Health regulation

Australian utilities strive to achieve the requirements of the *Australian Drinking Water Quality Guidelines* (ADWG) as specified by the National Health and Medical Research Council, and Australian Government institution, working with the Natural Resource Management Ministerial Council. The ADWG are designed to provide an authoritative reference to the Australian community and the water supply industry on what defines safe, good quality water, how it can be achieved and how it can be assured. The introduction to the latest iteration of the ADWG state that they are:

> "intended to provide a framework for good management of drinking water supplies that, if implemented, will assure safety at point of use. The ADWG have been developed after consideration of the best available scientific evidence. They are

designed to provide an authoritative reference on what defines safe, good quality water, how it can be achieved and how it can be assured. They are concerned both with safety from a health point of view and with aesthetic quality" (National Health and Medical Research Council, 2011).

The ADWG adopted a "catchment-to-tap" approach suggesting actions that should be taken at all points in the supply chain to achieve high quality water at the point of use. Note should be taken of the term "Guidelines" in the title of the ADWG; these are not mandated standards, or at least not mandated by the federal government. It is down to the Departments of Health and similar agencies (depending on the state or territory concerned) to determine how this guidance should be applied. Generally, health regulators specify (through regulation or in licences issued to water utilities) the standards that are to be met, taking into account local conditions, the existing state of the system to be regulated, economic circumstances, risks and consumers' willingness to pay for improvement.

The ADWG are an element of the National Water Quality Management strategy that includes guidelines for recreational waters, aquatic ecosystems, urban stormwater, recycled water, effluent management, sludge (biosolids) management, and trade waste management, among others, and component guidelines for particular industries (e.g. effluent from intensive piggeries). Each of these guidelines is reviewed on a rolling basis, with community input (see http://www.environment. gov.au/topics/water/water-quality/national-water-quality-management-strategy).

4.5.7 Diversification of supplies

One of the effects of the regulatory/institutional arrangements that have developed over the past 20 years is an increased focus on water use efficiency and diversification of supplies, although it must be noted that these developments have also been stimulated by very severe periods of drought, during which water restrictions were introduced in most jurisdictions. The setting of a full cost-reflective price for water has rationalised water use such that consumers have a tendency to use only the amount of water they're willing to pay for. Thus, water used per capita has declined significantly, as described above. The Water Services Association of Australia reports, for example, that while Sydney's population has grown by 1.2 million since 1974, total water use in the city has not increased at all (Water Services Association of Australia, 2009). While drought has now broken over much of the country and most water restrictions have been removed, water efficiency has been "hardwired" in the system as residential consumers have invested in water efficient appliances (see Water Efficiency Labelling and Standards Scheme, below) and industrial consumers have reformulated production processes to reduce water use or to recycle wastewater. Avoided water use represents a major new "water source" driven by the reforms introduced over the past two decades.

As accessibility of water has declined and utilities have become responsible for their own investment decisions, alternatives to traditional surface and groundwater systems have been developed. Investments in desalination have occurred in most capital cities, as has investment in recycled water systems, and water sensitive urban designs intended to retain and reuse water within the urban environment. Whereas most urban utilities formerly relied on one or two sources of water, their portfolios now commonly include surface waters, groundwater, stormwater and recycled water, desalination and avoided water use (demand management). Generally speaking, such diversification was not mandated by regulation, but was stimulated by the operational environment that focused utilities' attention on price, efficiency and returns on investment.

4.5.7.1 Water efficiency labelling and standards scheme

The Water Efficiency Labelling and Standards Scheme (WELS) is referred to in the National Water Initiative. It is an initiative that that requires certain products to be registered and labelled with their water efficiency in accordance with the standard set under the national Water Efficiency Labelling and Standards Act 2005. Under the scheme plumbing products (e.g. showers); sanitary ware (e.g. toilets); and whitegoods (e.g. dishwashers) are rated for their water efficiency. Consumers in the market for such products are provided with information showing the relative water efficiency of each device. Additionally, WELS sets some minimum standards for devices. Thus, for example, a toilet cannot be sold in Australia that uses more than 5.5 litres per flush. WELS has proven very valuable in aiding consumer purchasing decisions and reducing water consumed domestically. WELS has enforcement capacity, but generally relies on consumer education and developing sound working relationships with manufacturers to achieve its goals.

4.5.8 Consumer protection

As government-owned utilities are subject to the law just as private companies are, so too are utilities' customers protected by consumer law. Additionally, however, utilities have often developed of their own accord, or been required to develop under their operating licences, charters of consumer rights which set out the rights and responsibilities of consumers. For example, such a charter might specify the compensation available should a business experience a water service interruption of more than a limited period, or the pressure that will be maintained to a premises, say, 98% of the time. Furthermore, in many jurisdictions, recourse is available for dissatisfied customers to state Ombudsmen, either of a general nature, or specific to the water sector. Ombudsmen's offices can intervene to solve disputes or may direct the utility to make restitution to a customer.

4.6 ACHIEVEMENTS AND ROOM FOR IMPROVEMENT

The Australian water reform process has been quite radical. Rather than relying on a strict regulatory regime with the inefficiencies to which that is said to lead, market forces have been harnessed to create a competitive environment in which water use has been rationalised and utilities placed on a sound financial footing. While strict regulatory approaches do exist in many circumstances – utilities are, for example, responsible for meeting all legislative requirements, including those related to pollution control, health and corporate governance requirements – change has been driven primarily by structural/commercial reform. The results have been impressive, including very significant improvements in efficiency, large reductions in water use per capita and a reduction in debt. There is almost universal acceptance of value of the approach taken (see, for example, Victorian Competition and Efficiency Commission, 2011). The development of the national competition policy framework, and the water reforms that were responsive to that is seen as a seminal moment in modern Australian economic and natural resource policy. One of the most important features of the reform process was that it applied nationally; there are relatively few examples of such a concerted national effort.

There have, however, been some negatives, and activities that require further attention. These include:

- *Obscure objectives.* As described earlier, corporatised water utilities are generally granted a licence to operate and this usually includes the objectives shareholding governments expect utilities to achieve. There has been a tendency, however, for these documents to be extended to objectives beyond those that would normally be assigned to water service providers. That is, for example, they may include social policy objectives related to sustainability, equity, consumer protection, natural resource management, and others that are beyond those that would be required of a private company and which might compete with the primary objective of supplying fit-for-purpose water at an affordable price. The Australian government's Productivity Commission which recently held an inquiry into urban water management commented negatively on the tendency for operating licences to be used as a way of exerting government control over utilities (see Productivity Commission, 2011 for a more complete discussion of this issue).
- *Government involvement in decision-making.* This issue is not unrelated to that described above, but such involvement is exercised in ways other than through operating licences. Thus, for example, it was the New South Wales government that made the decision to invest in a desalination plant in Sydney, despite the point at which an independent review suggested the decision should be made not having been reached. The desalination plant investment was in the order of many billions of dollars. Had better risk management techniques been applied – as is likely would have been the case had government not been involved – the plant may not have been built, would have been of a smaller scale or would have been built later. As it is, the plant has never been required for water supply (although it is run to keep it in operational order). This pattern has been repeated in other capital cities.

- *Uncommercial dividend demands.* Commercial companies operating with independent boards would normally be expected to determine the dividend to be paid to shareholders. There are suggestions, however, that governments are too involved in this process, extracting dividends from utilities that are not commercial. This affects funds available to utilities for reinvestment and potentially increases debt levels.
- *Inconsistency in economic regulation.* Independent economic regulators such as the Essential Services Commission (Victoria) and the Independent Pricing and Regulatory Tribunal (New South Wales) are mature agencies that are rigorous in their approach. While there is a general tendency toward creation of such entities across the country, many are embryonic, with pricing decisions sometimes lacking transparency or subject to political interference.
- *Lack of progress on Water Plans.* Water plans affect rural systems more fundamentally than those in urban areas. Nevertheless they are relevant and important for all systems. However, as observed by the National Water Commission in all of its biennial reviews of progress on the National Water Initiative, governments have been very slow to initiate and complete Water Plans.
- *Lack of structural consistency.* The intent of the NWI has been to create commercially-focused utilities for urban areas across the country. This was achieved rapidly in some jurisdictions. Victoria, for example, has had in place water utilities for all areas across the state for many years, as has Western Australia and the Northern Territory. In other states, however, progress has been slow or inconsistent. Thus, while some entities in New South Wales have been corporatised for many years, for example, others – principally those managed by local governments rather than the state government – have languished. Often pricing, management and regulatory responsibilities in these areas remain unclear and there is evidence that the financial situation faced by some non-capital city utilities is weak.
- *Lack of competition.* Despite a desire to stimulate competition, competitive forces are still largely limited to benchmarking and contracting out services to the private sector. There is very little direct competition for the provision of water services in urban areas, although this situation is beginning to change.

4.7 FUTURE DIRECTIONS

The population of Australia continues to grow and, of course, its climate is as variable as ever. The process of water industry reform is not complete, nor is it ever likely to be. Catering to greater numbers of people, climate change, changes in the industrial bases of Australian industry, changes in consumer preferences, the need to ensure that water from all sources is safe and fit for purpose, and many other factors will each drive further changes. As a general comment, the direction of reform appears to be a tendency to harness market forces more completely, but this is not certain. Indeed, after decades of reform and the easing of drought, the Australian public's appetite for further reform may be limited, as may the willingness of its politicians to introduce further changes. Nevertheless, it is possible to cite the following as important trends, which may affect the nature of regulation of the water sector:

- *Integrated urban water management.* As water sources are diversified it becomes ever more important to integrate the various sources available for use so that multiple outcomes are satisfied. Objectives might include sustainability, energy use reduction, reliability, health protection, cost efficiency, and so on. The way in which various sources are used affects these outcomes. For example, the Water Services Association of Australia had observed that reduction in effluent discharge to the sewerage system arising from greater water use efficiency within the home has in some circumstances threatened the viability of recycling schemes which are dependent upon a regular supply of effluent for their operation (Water Services Association of Australia, 2009). In another instance a local government-based regulation mandating the installation of household scale rainwater tanks in a new developments reduced the demand for non-potable water such that a recycled water scheme's operation was threatened. An integrated approach to urban water management utilising diverse sources is needed. Where regulatory barriers exist to such integration they will need to be removed.

  More substantially, research attention is being increasingly directed to the development of water sensitive cities, in which the design and operation of integrated urban water systems is built into the fabric of cities. For example, urban water features could be created that treat stormwater sensitively using natural processes and which could also be a source of non-potable supply, a counter to urban heat-island effects or of aesthetic benefit. Presently many regulatory barriers exist to such developments. For example, fire protection requirements specify the use of potable water provided from a main of particular diameter and this limits system flexibility and increases costs. In fact, fire protection could be provided through other means (e.g. highly treated recycled effluent) just as safely. Furthermore, the way costs are apportioned in such systems is a constraint. For example, property developers may be charged the same costs whether they install traditional or innovative systems, even though the latter may be less costly to the utility. Finally, political exigencies have an effect on the way systems can be used. For example, in some jurisdictions there are politically inspired barriers to the transfer of water from rural to urban areas, or on the use of recycled water. In each of these instances regulatory reform may be required.

Brown, Keath and Wong (2009) suggested that modern cities are on (or could be placed on) a path from resource security, focused only on securing water supplies, to one in which multiple objectives – climate resilience, sustainability, amenity and others – could be achieved. They suggest we now have the potential to move from a "water cycle city" to a "water sensitive city" which is creative and adaptive. Figure 4.4 illustrates this concept.

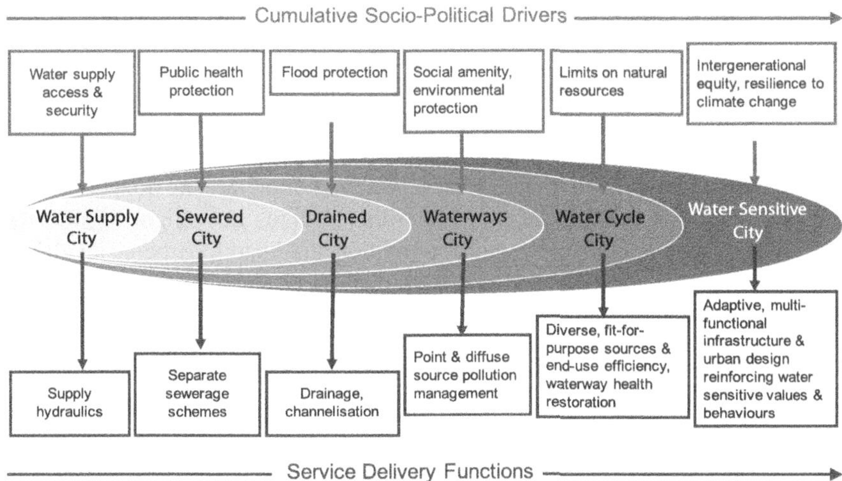

Figure 4.4 Pathway to water sensitive cities (Brown *et al.* 2008, p. 850).

* *Further price reform*. As indicated above, price reform is an on-going process. Several key issues have emerged however in the face of drought that require further consideration. For example, restrictions were put in place in most jurisdictions during the height of the drought limiting water use. Restrictions are blunt instruments that take no account of the preferences of consumers. For example, restrictions limited the watering of playing fields, but this also limited the opportunity for recreation. It is not clear that consumers valued recreation less than the price of water. There are also equity considerations. Thus, a person in a poorer neighbourhood is prevented from running a hose in the backyard in which children could play, but someone more privileged is not prevented from filling his or her swimming pool. It has been suggested that rather than impose restrictions, which have cost the country billions of dollars in lost productivity (Productivity Commission, 2011), it may be better to tie water price to availability. Such "scarcity pricing" would give consumers a choice in how they invested their money, and would provide flexibility in the allocation of water. There are several models that could be used. For example, a sliding scale could apply so that water becomes more expensive during drought, or consumers could choose to pay for a less for a less secure water supply which would be subject to restrictions in time of

drought, or more for a supply that not be subject to restrictions. While these are interesting notions, the price elasticity of demand for water (the extent to which a unit increase in price leads to a reduction in consumption) is currently understood to be low. Thus, scarcity pricing may have little effect on demand. This understanding deserves to be revisited in the light of recent drought experience.

- *Openness to competition.* One of the intentions of National Competition Policy generally and the national water reforms specifically has been to foster competition. Generally, this has been achieved though benchmarking, comparison and contracting out. However, there remains ambivalence to natural monopolies in some quarters, and arguments have been mounted that, in fact, water systems are no longer natural monopolies given the diversity of water sources that might be utilities. The opportunities to extend competition are therefore being explored, and legislative/regulatory change is occurring. For example, New South Wales has introduced the Water Industry Competition Act, which among other things, creates an opportunity for private firms to enter the market for water supply. Several organisations have already applied for licences, some to provide recycled water and one other to provide potable water within Sydney Water's area of operations. The effect of this legislation is receiving close scrutiny. While competition is favoured, there is also concern that private firms will be able to pick out the most profitable areas within the water system, leaving Sydney Water as a supplier of last resort. Others suggest that competition will reduce costs and produce significant consumer and environmental benefit.

4.8 CONCLUSION

The Australian water reform process has been transformational. Very large gains in efficiency of resource use (financial and other) have been achieved, and water demand has declined significantly. To a large extent these achievements have been market, rather than regulation, driven. While a strict regulatory environment exists, it applies equally to government-owned utilities and private companies. This, and the restructuring of utilities as companies with expectations of profitability and returns on investment, has stimulated an environment in which commercial rather than regulatory pressures have driven performance improvement.

Of course, the process has not been a perfect one. There are multiple governments involved and not all have responded as effectively to these drivers. Furthermore, much remains to be done as urban areas transition to water sensitive cities, and remaining reform processes are completed. There is always the risk, too, of backsliding. The new Australian government is, according to media reports, considering rolling the National Water Commission into the federal Department of Water, which would rob it of its independence and moral force. However, there is broad acceptance of the water reforms implemented to date, and so much of the process has been completed that it would be impossible to revert to previous practices even if government commitment to reform wanes.

4.9 REFERENCES

Australian Bureau of Statistics (2014). http://www.abs.gov.au/ausstats/abs@.nsf/Latest products/3218.0Main%20Features32012-13?opendocument&tabname=Summary&pr odno=3218.0&issue=2012-13&num=&view= (accessed 7 April 2014)

Brown R. R., Keath N. and Wong T. H. F. (2009). Urban water management in cities: historical, current and future regimes. *Water Science & Technology*, **59**(5), 847–855.

Bureau of Meteorology (2013). The National Water Account. http://www.bom.gov.au/water/ about/publications/document/InfoSheet_7.pdf (accessed 13 April 2014)

Council of Australian Governments (1994). The Council of Australian Governments' Water Reform Framework. http://www.environment.gov.au/resource/council-australian-govern ments-water-reform-framework (accessed 9 April 2014)

Council of Australian Governments (2004). Intergovernmental agreement on a National Water Initiative. http://nwc.gov.au/__data/assets/pdf_file/0008/24749/Intergovernmental-Agree ment-on-a-national-water-initiative.pdf (accessed 9 April 2014)

Council of Australian Governments (2010). National Water Initiative Pricing Principles. http://www.environment.gov.au/system/files/resources/34dbb722-2bfa-48ac-be7e- 4e7633c151ed/files/nwi-pricing-principles.pdf (accessed 12 April 2014)

Department of Environment, Water, Heritage and the Arts (2010). National Water Initiative Pricing Principles: Regulation Impact Statement. Australian Government, Canberra. http://www.environment.gov.au/system/files/resources/78883fe0-c5b6-4dec-9fac- 5b630d064f00/files/ris-nwi-pricing-principles.pdf (accessed 12 April 2014)

Environment.gov.au. (2014). National Water Initiative Water Plans. [online] Available at: https://www.environment.gov.au/water/australian-government-water-leadership/nwi/ status-incomplete (accessed 13 April 2014)

Hoang M., Bolto B., Haskard C., Barron O., Gray S. and Leslie G. (2009). Desalination in Australia. CSIRO, Australia

National Health and Medical Research Council & Natural Resource Management Ministerial Council (2011). Australian Drinking Water Guidelines 6. Australian Government, Canberra. https://www.nhmrc.gov.au/_files_nhmrc/publications/attach ments/eh52_aust_drinking_water_guidelines_update_131216.pdf (accessed 13 April 2014)

National Water Commission (2005a). Australia Water Resources, 2005: A Baseline Assessment of Water Resources for the National Water Initiative. http://www.water. gov.au/publications/AWR2005_Level_2_Report_May07.pdf (accessed 1 April 2014)

National Water Commission (2005b). Australian Water Resources 2005. http://www.water. gov.au/WaterUse/Waterusedbytheeconomy/index.aspx?Menu=Level1_4_2 (accessed 8 April 2014)

National Water Commission (2011a). The National Water Initiative – Securing Australia's Water Future: 2011 Assessment. National Water Commission, Canberra. http://www. nwc.gov.au/__data/assets/pdf_file/0018/8244/2011-BiennialAssessment-full_report. pdf (accessed 13 April 2014)

National Water Commission (2011b). Review of Pricing Reform in the Australian Water Sector. National Water Commission, Canberra. http://archive.nwc.gov.au/__data/assets/pdf_file/0015/18051/Chapters_3_-_5.pdf (accessed 9 April 2014)

National Water Commission (2012). National Water Planning Report Card. http://archive.nwc.gov.au/library/topic/planning/report-card (accessed 13 April 2014)

National Water Commission & Water Services Association of Australia (2013). National Performance Report 2012-13. Australian Government, Canberra. http://www.nwc.gov.au/publications/topic/nprs/national-performance-report-201213-urban-water-utilities (accessed 8 April 2014)

Productivity Commission (2011) Australia's Urban Water Sector (Vol. 1). Australian Government, Canberra. http://pc.gov.au/__data/assets/pdf_file/0017/113192/urban-water-volume1.pdf (accessed 10 April 2014)

Radcliffe J. C. (2006). Future directions for water recycling in Australia. *Desalination*, **187**(1), 77–87.

Tisdell, J., Ward, J. and Grudzinski, T. (2002). The Development of Water Reform in Australia. Cooperative Research Centre for Catchment Hydrology, Canberra. http://members.iinet.net.au/~jtisdell/utas_website/pdf/Tisdell_development.pdf (accessed 2 April 2014)

Victorian Competition and Efficiency Commission. (2011). Victoria's Productivity, Competitiveness and Participation: Interstate and International Comparisons. ACIL Tasman Pty Ltd, Melbourne. http://www.vcec.vic.gov.au/CA256EAF001C7B21/WebObj/ACILTasmanReportWord/$File/ACIL%20Tasman%20Report%20Word.doc (accessed 23 April 2014)

Water Services Association of Australia (2009). Vision for a Sustainable Urban Water Future: Position Paper No 3. https://www.wsaa.asn.au/Resources/Positions/Vision%20for%20a%20Sustainable%20Urban%20Water%20Future.pdf (accessed 14 April 2014)

Water Services Association of Australia (2012). Asset Management Performance Improvement Project. https://www.wsaa.asn.au/projects/Pages/Asset-Management-Performance-Improvement-Project.aspx#.U0lGC_l9KSo (accessed 12 April 2014)

Chapter 5

Regulation of water services in Denmark – a utility manager's perspective

Jens M. Prisum

CTO, BIOFOS, Copenhagen; *Board Member, Danish Water and Wastewater Association Board Member, Eureau, and Board Member, Water Supply and Sanitation Platform (WSSTP)*

5.1 INTRODUCTION

This paper explains a recent reform of the Danish Water Sector, which among other items includes the introduction of a price cap regulation setting a maximum for the tariff a water utility can charge consumers. The principles of the regulation are described and the consequences of the implementation are presented as seen from the point of view of a utility manager.

5.2 THE DANISH WATER SECTOR

The Danish Water Sector is very fragmented with more than 2,500 utility companies serving a population of about 5 million people. The companies are owned by municipalities or consumers. However most of the water supply, sewerage collection and treatment are performed by less than 200 mainly municipal owned companies, see Table 5.1.

The turnover in the Water Sector is about DKK 10 billion/year (1 Euro = 7.5 DKK).

Table 5.1 Structure of the danish water sector.

Water Supply	Sanitation
Approx. 2,300 companies, mainly small cooperatives. (10% of total supply)	100 Companies Owned by municipalities. (98% of supply)
200 larger companies serving cities and part of rural areas – primarily owned by municipalities. (90% of total supply)	

5.3 THE DANISH WATER SECTOR REFORM IN BRIEF

In 2003, The Danish Competition Authority estimated the annual potential for improving efficiency in the Water Sector to be 1.3 billion DKK. This led to a political debate about how to reform the water sector, the outcome of which was a new Water Sector Law in 2009.

The main purpose of the law is to improve efficiency of utilities in order to be able to reduce tariffs and finance future costs associated with implementation of the Water Framework Directive and Climate Change. The scope of the reform is defined in the Water Sector Law and supplementary Ordinances and Guidelines.

Basically, the law requires water supply and sanitation activities to be separated from the general municipal services and organised as a limited company owned by the municipality. (The municipality can sell the company, but as it is not allowed to make any profit, there are no buyers).

The law introduces a price cap regulation and the establishing of a Regulator, based in the Competition Authority.

The regulation includes utilities supplying a volume of 200,000 m³/year or more which corresponds to the larger companies in Table 5.1.

The law specifies that every year, the Regulator shall issue a maximum value for the tariff each utility can charge consumers. This value is determined from a comparison of performance in an annual benchmarking exercise, supplemented by an evaluation of inflation and general productivity development in society. The results of the Water Sector Law will be evaluated in 2014.

The environmental performance of the utilities is – as previously – monitored by the Ministry of Environment.

Apart from the regulatory "top down" benchmarking, the Danish water and wastewater utilities have for more than 10 years worked actively on improving cost efficiency through a voluntary "bottom up" benchmarking system, developed and operated by their own association, Danish Water and Wastewater Association, DANVA.

The principles of the price cap regulation and the DANVA benchmarking system are described in sections 5.4 and 5.5.

5.4 PRICE CAP REGULATION ACCORDING TO THE WATER SECTOR LAW

The law requires that by 1 January 2010, municipally-owned utilities should be separated from the general municipal services and organised as a limited company. These utilities are allowed to charge the consumers up to a price cap determined as follows:

Maximum tariff = Price cap for OPEX + Additional costs the utility cannot control such as taxes and costs associated with environmental targets and service levels set by public authorities + Price cap for CAPEX.

5.4.1 OPEX

The basis for the regulation is the average annual OPEX in 2003–2005. From this basis, each year the efficiency demand for a utility is determined by adding inflation and deducting a general estimate of the development of production efficiency in society. Finally, an individual efficiency demand is set based on a benchmarking of utilities.

The benchmark model is based on the principle of Data Envelope Analysis (DEA). DEA estimates the cost level an efficient company should be able to achieve in a particular market. This is done by using cost drivers that reflects the costs of different activities carried out when producing water or treating wastewater. An example of a cost driver for the supply of drinking water could be amount of water produced, length of distribution network etc.

The benchmarking defines the efficiency potential for each utility by comparison with the most efficient peers. From this potential, the annual individual efficiency improvement demand for a utility is determined after allowing for 20% uncertainty. Also individual conditions for each utility can be included, provided the utility can document this. Finally, no utility can have an annual efficiency demand of more than 5%. If the utility spends less on OPEX that what is allowed by the price cap, this can be used for financing CAPEX. No profit or return on equity is allowed.

As an example, the efficiency potential for selected larger utilities in 2013 is shown in Figure 5.1. As shown, for some utilities an efficiency potential of 20% or more is estimated by the benchmark model. Generally, the Regulator assumes that the efficiency potential can be realised within 4–5 years.

However, as the annual efficiency demand reflected in the price cap cannot exceed 5%, the individual efficiency demands do not reflect the potential as seen in the Figure.

         Regulation of Urban Water Services

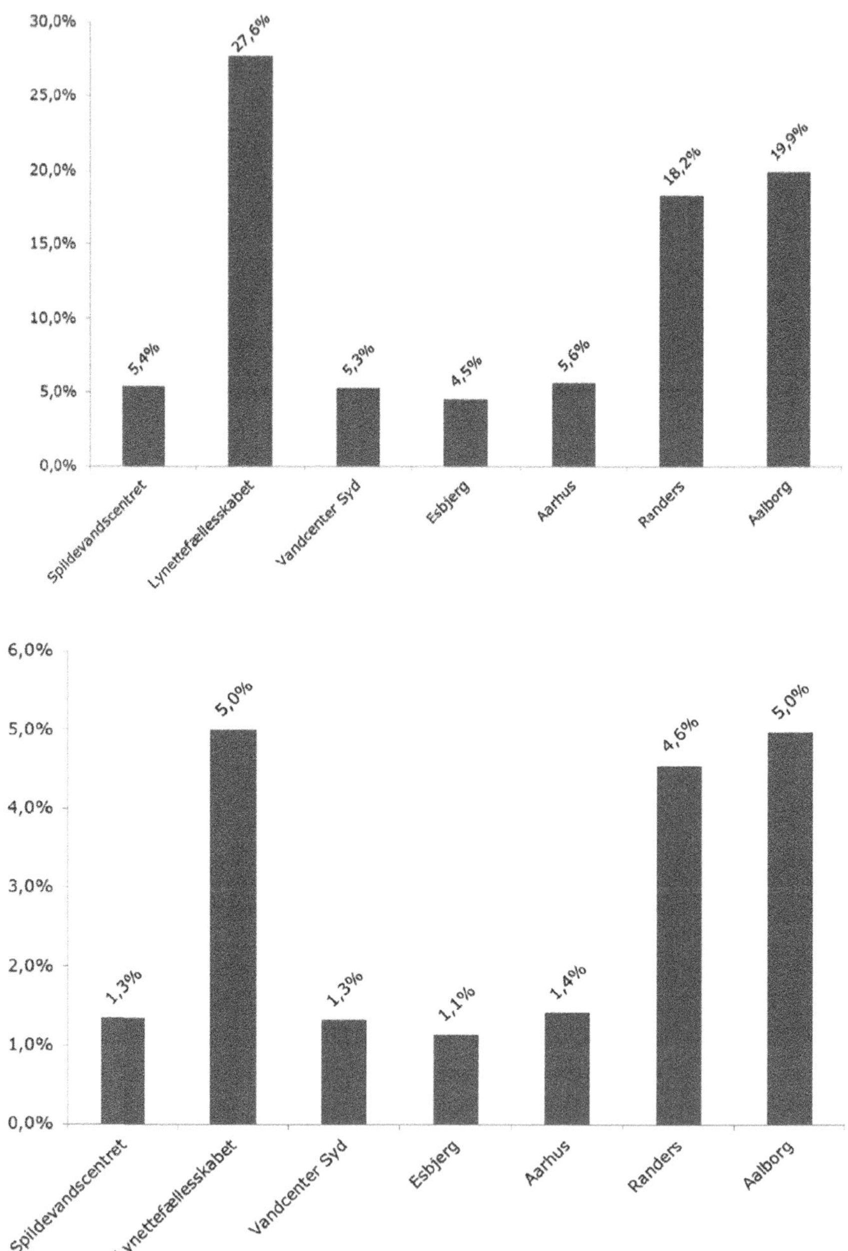

Figure 5.1 Selected utilities 2013, Efficiency potential (upper figure) and individual demand (lower figure).

5.4.2 CAPEX

For CAPEX a price cap is set corresponding to the estimated depreciation of assets.

No efficiency demand is introduced. The reason for this is that at the time the regulation was introduced heavy investments were needed for renovation and expansion of sewers to cope with the effects of climate change.

In order to align the starting point for each utility, for regulation purposes, the value of the assets for each company were set based on a "cost catalogue" developed by the Regulator. From this catalogue, the value of an asset, e.g. a wastewater treatment plant, could be determined based on capacity as population equivalents. In order to simplify bookkeeping, and if accepted by the auditor, a utility could choose to replace the assets noted in the actual balance sheet with the value determined from the cost catalogue.

From this starting point a price cap on CAPEX is set as the estimated annual depreciation of assets. If investments require more liquidity, this must be financed by mortgage or other loans. Interests are included in the OPEX price cap and repayment of loans in the CAPEX price cap.

A surplus on CAPEX price cap cannot be used for financing OPEX, but will be deducted from the CAPEX of the coming years. To illustrate the mechanism, the price cap for one of the BIOFOS subsidiary companies is shown in Table 5.2. As seen from the table, the individual efficiency demand is targeting about 50% of the total maximum price.

Table 5.2 Price cap 2014, BIOFOS – spildevandscenter avedore.

OPEX cap within efficiency regime	4.47 DKK/m^3
Environmental and/or service level goals	0.77 DKK/m^3
CAPEX cap	3.68 DKK/m^3
Maximum price:	8.92 DKK/m^3

Note: m^3 is drinking water supplied

5.5 THE DANVA BENCHMARKING SYSTEM

Apart from the benchmarking system used to set the price cap, the Water Sector Law requires utilities to perform an annual process benchmarking exercise in order to become more efficient. For this purpose, most utilities use the benchmarking system developed by DANVA. This system has been in operation for more than 10 years. Utilities themselves submit data to the system via the internet from which unit costs of different processes are determined and made available to participants in tables and graphs as shown in Figure 5.2. Apart from unit costs the DANVA system also allows for a comparison between other Key Performance Indicators (KPIs) such as effluent values and removal rates, see Figures 5.3 and 5.4.

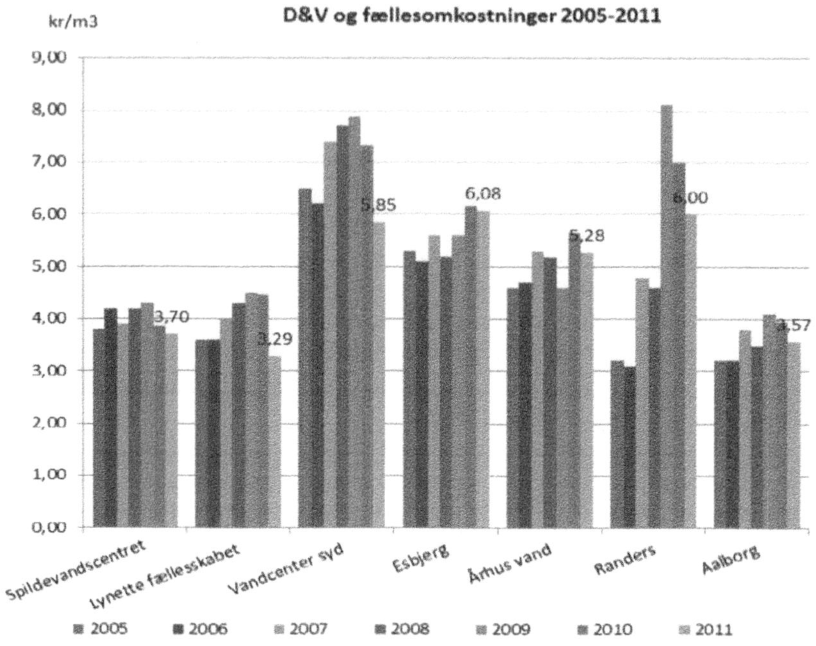

Figure 5.2 Selected utilities. OPEX for wastewater treatment. 2005–2011. DKK/m^3 of drinking water supplied.

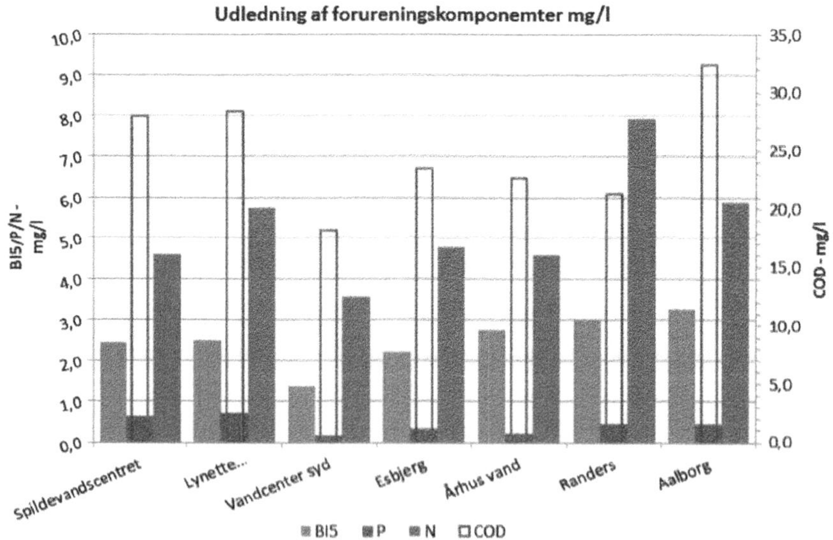

Figure 5.3 Selected utilities. 2011, Wastewater treatment. Effluent values mg/l.

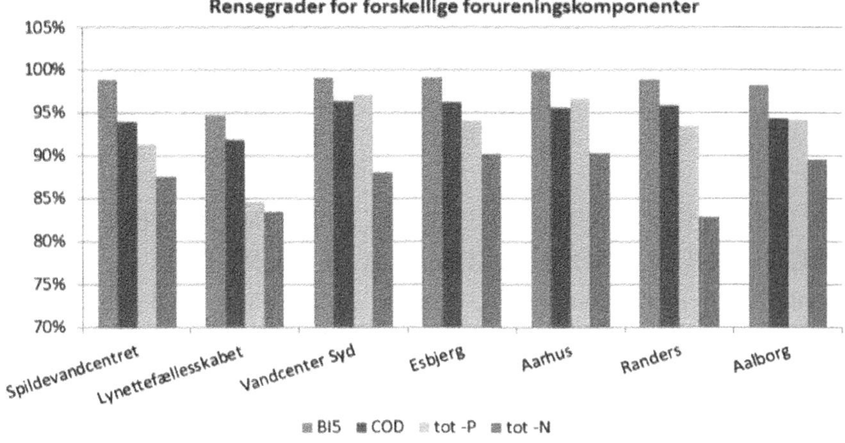

Figure 5.4 Selected utilities, 2011. Wastewater treatment. Removal rates %.

5.6 COMPARISON OF BENCHMARKING SYSTEMS

The DANVA and Regulatory benchmarking systems are compared in Table 5.3.

Table 5.3 Comparison of benchmarking systems.

DANVA	Regulator
• Ranks utilities based on KPI s	• BM ranks utilities for price cap determination
• Developed by practitioners for practitioners	• Developed by economists for economists
• Reflects business processes in utility	• Reflects theoretical costs drivers
• Basis for exchange of experience	• Exchange of experience mainly focuses on finding loopholes
• Results are made public	• Results are made public
• Annual exercise	• Annual exercise
• Modest data control (may be reflected in data quality)	• Lot of control procedures (and associated costs, auditors)
• Results are understandable for managers and practitioners	• Only specialists can explain results (CEO cannot explain the accounts)
• Little focus from Management	• High focus from Management

The major difference is that the DANVA system is developed by practitioners and focuses on Key Performance Indicators for business processes as well as environmental and service level parameters. The Regulator's system focuses on economy and efficiency and includes a lot of control procedures and associated costs from auditing and documentation. Finally, management tends to focus on the regulator's system as this has important financial consequences for the utility.

In the long term, however a KPI system such as DANVA is needed as a tool when management has to identify where performance can be improved.

5.7 A UTILITY MANAGER'S EVALUATION OF THE REGULATION AND BENCHMARK IN GENERAL

So far this paper has focused on the technical aspects of regulation of the Danish water Sector. In this section some more personal observations of a utility manager are presented. These are clearly subjective and reflect the opinion of the author only.

5.7.1 General observations

When introduced 10 years ago the DANVA system was generally found useful for improving efficiency by informal exchange of experience. Participants in such activities were most often technicians and not managers. Very few used the system as a base for improving financial performance and by many the main purpose of the system was seen as a way to demonstrate sector willingness to become more efficient and thereby avoid regulation.

This attitude generally changed and managers started to consider how they could use benchmarking as a tool for improving performance of the organisation. However, by then the price cap regulation was introduced and the entire focus turned to that as this is extremely important to the utilities.

Compared to the DANVA system where KPIs directly reflecting business processes are compared, the DEA analysis is a tool which is understood by specialists only. For smaller utilities in particular, which cannot afford such expertise within their organisation, this makes the results difficult to interpret for management. Thus a large debate has been going on since the regulation was introduced and is still going on, as to whether the results are correct and reflect the special conditions in "my utility". Great efforts were (and are still) spent on how to find loopholes in the regulation and get a higher price cap.

Other particularities have come out of the fact that OPEX is regulated but CAPEX is not. This has in some cases led to investments which from a normal financial evaluation were not sound.

5.7.2 Specific observations

5.7.2.1 Keep it simple

When introducing a regulation, make it transparent and as simple as possible in order that the management of the regulated utilities can actually understand the principles behind the regulation. Many Danish utility managers now find it difficult to explain the accounts to the Board of Management. Regulation should be set up in a way that does not motivate speculation in finding loopholes.

5.7.2.2 Include parameters other than financial

Regulation based on financial performance only does not reflect the performance of a utility with respect to the many demands from consumers and the general public. KPIs such as Consumer Service performance, i.e. number of Complaints and Breakages, and Environmental performance, such as treatment efficiency and carbon footprint, should also be reflected in the regulation.

5.7.2.3 Follow general principles of financial management

Make sure that regulation does not work in a way that turns normal methods of business management upside down, e.g. by favouring investments when they are not feasible from a normal investment analysis.

5.7.2.4 Allow for time to plan and implement

Annual determination of a price cap does not allow management to plan and implement. Any organisation will need time to identify and implement how to become more efficient. Thus a price cap reflecting a longer period – say 3 years – will be a better approach.

5.7.2.5 Utilities should develop a proactive attitude towards regulation

Although the regulation is still debated, in Denmark more and more utility managers see this as an important driver for the development of the sector in general and their organisation in particular.

As profit is not allowed, and thereby lacking this parameter to demonstrate performance, utilities should use the results of regulation and DANVA benchmarking to visualise their performance to interest groups such as consumers, owner, board, authorities, non-governmental organisations (NGOs), general public and colleagues.

5.7.2.6 The water sector reform is accepted

Utility managers have accepted the reform of the Danish Water Sector. In particular, the way utilities have been made more independent as limited companies has freed a lot of momentum and has put a higher focus on management and performance rather than technology.

Few want the former system back, but most hope for a simpler system in line with the observations above.

To conclude, the author's personal experience of the regulation introduced by the Danish water Sector Reform can be summarised as follows:

- Regulation and top down BM have changed the focus of utilities from technology to management and performance.
- BM must be combined with Regulation to be taken seriously.
- Bottom up BM, such as DANVA's can be used to identify possible improvements.
- Regulation and BM should be so simple and transparent that managers can understand it.
- Regulation / BM period should be at least 3 years in order to plan and execute.
- Control is necessary, but must be kept at a minimum.
- Regulator and utilities should be in continuous dialogue on how to improve and simplify the system.
- Models and figures are necessary, but common sense and sector knowledge is a prerequisite.
- Transparency, internally and externally is important. We are a monopoly!

Chapter 6

Experiences and conclusions about regulation in Latin America and the Caribbean

Andrei Jouravlev

6.1 INTRODUCTION

As a result of the recession in the eighties, the role of the State in the economy in general in Latin America and the Caribbean, and in public utility services including drinking water and sanitation, has changed radically (Jouravlev, 2004). The objective of this change was to reduce and redirect public spending in a context of tax austerity measures and to increase the efficiency in public utility service provision. One of the main results of these reforms was that the functions of the State were transferred from direct exploitation of water, execution and operation of public works and direct provision of public utility services, to that of regulator, control and promotion of activities by third parties, whether autonomous public bodies, local governments or private companies (CEPAL, 1992).

As a result of these policies nearly all the countries in the region have reformed the institutional structure of the drinking water and sanitation sector (Jouravlev, 2004). These reforms invariably involved an explicit institutional separation between the functions of defining sectoral policies, hierarchy and strategic planning; economic regulation, supervision and control of the companies; service provision and operation of the infrastructures.

In most of these countries today, these three functions are assigned to different organisations and the rights and obligations are clearly defined. This functional separation, which is the cornerstone of the reform process in the sector, and which can be seen in general in all public utility services, represents major institutional progress. Regional experience suggests that dividing up the functions is indispensable in any cases when it has been decided to privatise service provision, but it is also highly recommended even when state or municipal control is maintained over them (CEPAL, 2000).

In 70% of the countries in the region, autonomous economic regulation bodies have been created for drinking water and sanitation services, either at national level (in nearly all cases), sub-national level (in federal countries) and even at municipal level. In most cases the intention is to regulate the public utility services sector separately, and therefore specialist entities have been created to regulate the drinking water and sanitation services, whereas in others, normally the smaller countries, there are proposals for one single regulating body for all public utility services (CEPAL, 1998a, 1998b).

In 2001 in response to the need to drive integration and cooperation in sectoral regulation in the region, the "*Asociación de Entes Reguladores de los Servicios de Agua Potable y Saneamiento de las Américas*" (ADERASA) was established [Association of Drinking Water and Sanitation Service Regulators of the Americas]. Among ASERASA's priorities, cooperation in five relevant subjects for regulation in the drinking water and sanitation sector in the region are highlighted: competence by benchmarking, small operators (efficient regulatory instruments for small operators), public providers (regulation specifics under the public service provision model with state and municipal owned operators), "green" regulation (regulatory tools that could achieve inclusion of environmental costs in drinking water tariffs), and regulatory accounting.

Although the institutional design of regulating bodies has sought to provide them with a higher degree of autonomy and independence, mainly against political interference, in practice weak organisations have been created, without any real authority, with extremely limited degrees of discretion and inadequate mechanisms for resolving conflicts that undermine the regulatory function of the State (Jouravlev, 2004). In many cases these organisations are subject to *ad hoc* intervention of the executive power, usually in favour of providers. Moreover, there are usually conflicts of competence with local governments, unstable managerial posts, low budgets that do not allow them to carry out their work effectively and limited legal capacities to perform their functions.

More recently, several countries have strengthened their regulatory systems and organisations, whereas in others, due to withdrawal by private international operators and providers returning to state ownership, the regulatory bodies have been weakened, particularly in those countries where there have been serious conflicts with private investors (Ducci, 2007). This weakening, and even questioning of the role of regulation, has arisen because between governments and consumers there is a feeling that the organisations did not properly fulfil the role they were assigned, although in many cases it has been admitted that they did not have sufficient resources or autonomy to perform their tasks.

6.2 WHAT IS REGULATION?

Drinking water and sanitation services are classic examples of natural local monopoly. A natural monopoly is an activity in which, because of the intrinsic technological characteristics involved, the total production costs are lower when the service is provided by one single service provider than when the service is divided up between two or more operators. Consequently, bringing in another service provider is not profitable and the fact that the service in a specific geographical region is in the hands of one single provider is more efficient.

With regard to natural monopolies, governments are faced with two fundamental alternatives when providing citizens with a service: state owned, as has historically been the case in most countries, or regulation of privately owned monopolies. If state ownership is abandoned, Governments need to intervene as regulators in order to encourage productive efficiency and in assignations in view of the lack of competition in a branch of activity that is naturally monopolistic.

Regulation has the aim of reproducing the results that would be achieved through productive efficiency and efficiency of assignations in a competitive market system. This is known as the principle of subrogation of markets and obeys a traditional view of regulation applied to privately owned providers. According to this idea, in activities where a natural monopoly is inherent, the regulator should act as a market substitute, in an attempt to force the service provider to essentially behave as it would if there were no regulations but as if it were subject to the force of free competition.

In Latin American and Caribbean countries, drafting regulatory frameworks and creating responsible organisations was a comprehensive part of the reforms of the nineties, which in many cases had the principal aim of attracting investment and private management capacity for the drinking water and sanitation industry. However, incorporation of the private sector was not particularly successful and did not last long because of the structural limitations of national economies, changes in corporate strategies by transnational companies, the presence of conflicts linked to tariff increases to solve the planned investments and changes in the global economic scenarios (Lentini & Ferro, 2014). This led to withdrawal by private service providers in the region from the early 2000s and a gradual return to state and municipal owned service provision (Ducci, 2007), with the significant exception of Chile. Nevertheless, the regulatory organisations did last and in most cases maintained the legal frameworks originally designed to regulate private activity but which ended up being applied to public operators of state or municipal owned services.

One of the main challenges the countries in the region face today is that on the one hand service provision is characterised by the prominence of public entities, but on the other, regulatory frameworks have been designed with the specific objective of regulating private service providers by means of economic and financial incentives. The problem is that these incentives are not necessarily effective, and in some cases, in the framework of public service provision, can be counter-productive.

For example, particularly at municipal level it is common for service providers, or their institutional proprietors, to be reluctant to adjust their tariffs to self-financing levels, for political reasons (Ducci & Krause, 2012), as they prefer to depend on budget transfers from other government levels (normally from central governments). In other cases, they simply ignore regulatory mandates, because of their negotiation power or support from other executive powers. Similar conflicts frequently arise as a result of municipal autonomy, which is often made worse because of the fragmentation of service providers. Hence, many operators are forced to work in an environment of limited resources meaning they are unable to comply with regulatory mandates. Most service providers do not manage to be financially self-sufficient whereas the decisions about assigning budget resources and permitted debt levels do not normally depend on the regulator or the provider, but rather the tax or finance authorities.

The subject of how to regulate a state or municipal owned service provider is open to speculation (Solanes, 2007). There can be no doubt that the normal controls of a public company are relevant, but neither can there be any doubt, owing to specificity, that providing drinking water and sanitation services requires controls that have been specifically defined for this activity. The challenges were analysed at an Experts' Meeting about Tariff and Regulatory Policies within the framework of the Millennium Development Goals and the Human Right to Water and Sanitation (Santiago de Chile, 8 July 2013) (Lentini & Ferro, 2014). This is also one of the priority cooperation subjects in the ADERASA framework.

6.3 CONTRACTS AS A REGULATION METHOD

During the first half of the nineties there was a lot of enthusiasm for privatising companies in the sector and nearly all the governments of the countries in the region adopted ambitious plans (CEPAL, 1998a, 1998b). Nevertheless, not many of these plans actually materialised.

Owing to the non-existence of regulatory frameworks or the organisations in charge of applying them, or the fact that they had not been consolidated, and the fact that private participation had to be achieved at any cost (Solanes, 2002), several countries decided to include private capital and management through different types of contracting systems, mainly concessions for major cities and BOT (building, operation and transfer) in the case of provided water treatment and desalinated seawater systems, with regulation through contracts predominating over a general law and autonomous agency. In some cases, moreover, regulation bodies were created (or control of contract application), sometimes at local level, either afterwards or at the same time as the private sector was brought into the industry, without these bodies having been previously consolidated.

When attempting to bring in sustainable private investment, the use of contracts as the main regulation method was not as successful as expected. Neither did it guarantee the efficiency of the services provided. In fact quite the opposite was true, as many conflicts arose leading to termination of most of these contracts.

According to regional experience, the main limitations this approach involves are as follows:

Public tenders may not be competitive, owing to the limited number of competitors and collusion between them. It is estimated that in 60% of the cases where there has been private participation in drinking water and sanitation services in the region, the "competition" was limited to one or two companies (Foster, 2001).

Opportunistic conduct after public tenders. Once contracts have been adjudicated, replacing the operator is difficult and expensive. There is a trend to submit speculative or overly optimistic bids and then try to renegotiate them later on.

Problems concerning specifications, supervision and application of contracts. Perhaps one of the most significant limitations of this approach arises when it is acknowledged, in a world of constant evolution, that the contractual conditions need to be amended as time goes by. It is estimated that more than 70% of the drinking water and sanitation service provision contracts have been renegotiated in the region (Estache *et al.* 2003), normally through conflictive processes which are sometimes not at all transparent. As expected, service providers are reluctant to reduce tariffs, and governments are reluctant to increase them. In several cases the contracts lacked any provisions for the administrative machinery to be able to control and supervise the service provider, and to verify compliance with the concession conditions.

Problems when terminating a contract with asset valuation and transfer. If operators expect the investments they make in the contract validity period could be undervalued at the time of termination, their incentive to invest in new assets and maintain the existing ones will be correlatively low. Consequently, there is a danger of operators wearing out the infrastructure towards the end of the validity period or building infrastructure with scheduled obsolescence in line with the transfer calendar. Moreover, it is possible that at the end of the contractual period, its strategic advantages have reached an extent, from operating experience and construction of the system, where potential competitors will refrain from submitting competitive bids.

On the other hand, the important lessons learnt through experience by the countries in the region during the nineties with the use of contracts as a regulatory method are as follows:

Contracts, whether the service providers are private, public or mixed, do not dispense with the need to regulate them, on the basis of information systems fed by thorough, standardised regulatory accounting (Vergès, 2010).

The definition of a regulatory framework, and the institutional design and implementation of the competent bodies, must precede the privatisation process (CEPAL, 2000). If this is not the case, reforms can be unstable, leading to unjustified equity transfers and revenue, sometimes for very high amounts which does not guarantee the efficiency of the provided service nor does it attract investment in the sector.

The enormous public interest related to the services and the need to settle intrinsically conflictive, complex matters justify the need for the *regulatory framework to be drafted in a law rather than a contract* (Jouravlev & Solanes, 2009). This approach has certain advantages in terms of the solidity of the legal structure of the system, the scope, seriousness and depth of the legislative debate and the visibility of the arguments and influences by the different political forces and interest groups.

The desirable characteristics for institutional design of regulating bodies is for them to be independent to a certain extent in order to avoid interference, particularly executive interference. At the same time, they should be accountable before legislative power (Solanes, 2007). Nevertheless, rather than simply depending on the letter of the law, this independence is more dependent on the governability or political culture of a society. Independence is fundamentally understood in the sense that decisions should only be appealable before the Courts, thus eliminating the possibility of administrative appeals which favour capture by the industry. Moreover, in many cases, the holders are forbidden from exercising political activity and stability in their posts.

6.3.1 Information access

The regulatory frameworks initially adopted by the countries in Latin America and the Caribbean for the drinking water and sanitation industries were weak, particularly when compared to the practices established in countries with more experience in this field (Solanes, 1999; Jouravlev, 2004; CEPAL, 2000, 2005). A critical matter was that some of these frameworks did not guarantee suitable access by regulators to the information they needed in order to carry out their functions properly, and furthermore, some of the design specifications (i.e. the way documents were submitted) exacerbated the information advantage by the regulated companies (Jouravlev, 2003). We should not forget that the most noteworthy characteristic of the modern economic regulation theory is the central role that information plays. From this perspective, regulation is fundamentally understood as a control problem within an asymmetrical framework of information between the regulator and the regulated companies (Laffont, 1994).

It was initially believed in some cases that modern regulatory instrument (such as price capping regulation, as in Denmark described previously) regulators would be able to obtain relatively limited, simple information about the costs and demand, would not have the need to measure the tariff base or the profitability rate and neither would they have to assign common costs, and in any case the use of contracts would drastically reduce the requirements for access to information.

Experience in concession of drinking water and sanitation services in Buenos Aires, Argentina, is very enlightening in this regard:

"Contracts meant that having established the goals in the drinking water and drainage supply services calls for bids, the physical and chemical properties, pressure, unitary production costs, distribution and availability, etc., the function of regulators would be negligible: limited to controlling that everything was performed according to the contract. In order to do so, there was an auxiliary for the specialist technical auditor and a financial auditor of acknowledged international prestige.
Nevertheless, after 1995 some relevant questions arose: Even though such and such works were not done, what was the current value of them? How much did the concessionaire save in operating costs by not doing them? How much income was lost through not receiving payment of invoices for this work and the boycott? Boycott is understood as the abnormal situation that was established. On the one hand the company stopped certain works, and on the other the users boycotted the concessionaire by not paying invoices. However these items (operating costs, investments to be made and income not received) were included in the equation at the time the rates were determined, the obvious question then arose: Should we reduce, maintain or raise the tariffs? By how much? The entire process has proved to be tremendously confusing, not transparent and very unpleasant. What was expected to be avoided through a call for bids for a fixed technical offer, at the best tariff, ended up in a worst case scenario, owing to the changes that reality itself brought to the originally forecast plans" (Dupré & Lentini, 2000).

These problems have forced many countries in the region to strengthen and perfect their regulatory frameworks in four principal areas:

Procedures to guarantee access to the service providers' internal information.
Firstly, it is essential for regulators to properly define what they need to know and how they want it to be delivered. Regulated companies usually have the legal obligation of regularly and permanently submitting accurate, truthful, relevant information covering a broad scope, according to established requirements, presentation formats, production criteria, itemisation, definitions, accounting standards and deadlines fixed by regulators. They are also obliged to supply any information requested by the regulators extraordinarily. It is indispensable to ensure that regulators have all the operating attributes to permit effectively obtaining the information they need: power to access and inspect facilities, inspect accounting and legal books, reports, equipment and installations of the service providers, and request sworn affidavits about information deemed to be relevant. The truthfulness of the furnished information is usually verified through technical and accounting audits by companies specifically contracted by the regulated companies, after previous certification by the regulating body. Auditors must be independent from the regulated companies and must provide the regulator

with a professional opinion based on commonly accepted standards and the
regulators' rules.

Regulatory accounting (Lentini, 2009). Regulators cannot carry out their
role effectively if they do not have the necessary attributes to define the
accounting system to be used by companies under their jurisdiction. When
introducing regulatory accounting, the following is normally sought:
(i) making information available about the income and costs compiled
in a standardised, consistent manner; (ii) access to information from
the different service providers obeying the same definition of criteria,
between the different providers and lasting over time; (iii) in the case
of companies who provide regulated services and other activities not
subject to the same regulation, providing information about the income
and costs specifically related to the regulated activity; (iv) itemising the
costs at the level required for regulation purposes (by activity, stage and
infrastructure installation); and (v) correctly recording certain types of
costs (for example, replacement costs separate from maintenance costs)
(Rodríguez, 2002).

Control of transfer prices (Hantke-Domas, 2011). As regulated companies
are also usually engaged in other businesses not subject to regulation (or
that are regulated under two different regulatory jurisdictions), regulators
must have access to the activity related to these activities, since otherwise
the transfer prices could be used to avoid economic regulation and to
support antitrust conduct in other markets. The same considerations are
applicable to transactions with linked concerns, even if the latter are not
regulated.

Competition by benchmarking (Ferro *et al.* 2011). This is based on the foregoing
regulatory tools, promoting indirect competition between companies
in the same business activity, but who operate in different geographical
areas, making the rewards granted to one of the companies dependent
on its own performance and on that of other service providers. The main
difficulties involved in implementing this instrument are related to the fact
that in practice there are always significant economic differences between
the services provided in different geographical areas and it is enormously
difficult to eliminate the influences of exogenous factors specific to local
conditions from the local costs, even if sophisticated statistical procedures
are employed.

6.3.2 Financial sustainability

One of the pillars of regulation is the need to generate sufficient incentives so that
the service provider charges a tariff that covers efficient costs (operational and
capital) so that it is financially sustainable (Hantke-Domas & Jouravlev, 2011).
Despite the fact that this simple rule in a region is more the exception rather than

general practice, in the last decade major progress has been made. There is an increasing trend, particularly with the bigger service providers, to cover operating costs plus depreciation through tariff revenue (Lentini & Ferro, 2014).

Optimum financing is achieved when it is completely covered by the users themselves (Hantke-Domas & Jouravlev, 2011). By creating a direct relationship between income and services provided (attended customers and supplied volumes, collected and treated), financial self-sufficiency generates incentives for better business efficiency (control of income and costs). Moreover, there is incentive for consumers to consume the services according to the valuation each of them makes of the services. Likewise, financial self-sufficiency reduces pressure on public budgets, meaning these amounts can be assigned to other areas of social welfare, and consequently reduce the space for political interference in the service provider's internal management, and also permits the service to be permanently, continuously financed, which reduces the vulnerability to economic swings which otherwise would depend on the budget located from the public treasury.

Optimum financial self-sufficiency is not a goal that can be achieved in the short- or mid-term. Obviously some will achieve it earlier than others, but all should strive towards this goal. All told, this task can be affronted if State policies are pursued that systematically acknowledge the milestones that need to be reached in order to reach the goal (Hantke-Domas & Jouravlev, 2011).

The first task is to identify the deficit and strongly invest in the infrastructure required to provide the service, on the understanding that the bulk of the investment, particularly in major works and networks, should be footed by public funds (Hantke-Domas & Jouravlev, 2011). This is a highly complex process since the total investment amounts are huge, and therefore effective prioritising should be guaranteed. Likewise, control over the investment must be guaranteed, since the high sums of money involved can be lost in transaction costs, corruption and capture or objectives external to the sector. Furthermore, an investment plan needs to be established containing the priority criteria for the use of public resources and even securing specific funds for specific works so that compliance can be controlled, at least by the tax authorities, in terms of execution, and by the regulator in terms of quality goals. Similarly, debt levels should not be overlooked, since high debt levels, particularly when they are external, can affect the financial stability of the service provider.

Another task should be instilling a payment culture, and efficient use of water, which must be associated with a tangible improvement in the service (Hantke-Domas & Jouravlev, 2011). The user charging service must be carefully designed, as it must be estimated in accordance with their financial capability; and charging should be made gradually over time in order to firstly cover operating costs and eventually recovering capital costs. This is the task that could extend over many years and must necessarily go hand in hand with public grant funds in the form of subsidies, particularly for low income groups. The objective will be for users to contribute to financing the operational costs and the investment, but even in this case some degree of subsidy must be maintained. The entire process must be controlled by a regulator capable of guaranteeing that the tariffs charged to users are in line with the costs necessary to provide the service at all times.

The importance of achieving the goal of financial self-sufficiency on the political agenda is a necessary condition for compliance, particularly in countries with major infrastructure deficits (Hantke-Domas & Jouravlev, 2011). This is of crucial importance since behind all lobbies for reforms and improvements in the sector there must be a motivating driving force which is none other than political will. Nevertheless, for this to actually materialise and last in time, a regulatory framework is required and institutions that implement those reforms in a sustainable manner. In terms of sustainability the main requirement stems from the need to impose the obligation of efficiency on service providers, so that a quality service is provided at the minimum cost in a sustainable manner.

For the purposes of determining efficient costs, a clear alignment stems from the need for service providers and the regulator to be supported by tariff surveys that permit estimating the expected demand, determining the infrastructure replacement requirements and any new work, and the operational costs (Hantke-Domas & Jouravlev, 2011). Once the budget requirements have been established, the exact income the entity will require can be calculated. Being supported by tariff surveys, beyond the convenience of having a planning tool, consists of generating an area of transparency and allows identifying when the decisions to reduce tariffs are not due to sustainability criteria, but rather to other considerations.

Once the costs have been defined, the required sources of income can be determined, which can basically come from two sources: the public treasury, i.e. taxpayers, or users (Hantke-Domas & Jouravlev, 2011). The former are from transferred budget funds at government level, and the latter through tariffs. Having said that, in reality it is always much more complex to define than is apparent in the theoretical models, although the truth is that there are no real cases where all income is from tariffs or from the public treasury. Nevertheless, it is advisable for a major part to be financed through tariffs rather than from subsidies. If water is considered from an environmental perspective with growing relative scarcity, individual payment is the most direct mechanism to increase awareness about how an irrational use of the resource can have undesirable effects that need to be accounted for.

The division between public financing and financial management by service providers is essential (Hantke-Domas & Jouravlev, 2011). Neither of the two should be confused. This means that the financial manager of the service provider should promote improved efficiency in processes, through good performance conduct, in order to obtain the desired financing with the revenue that can be collected from users. State or municipal contributions, on the other hand, tend to reduce the incentive by operators to collect funds, which is not conducive to improved management unless those funds are associated with or conditioned by improved efficiency. All said, public investments must always be accounted for in the service providers' balance statements in order to correctly calculate the operating, maintenance and asset replacement needs.

6.4 HORIZONTAL STRUCTURE OF THE SECTOR

In many countries in the region, towns have been traditionally responsible for providing drinking water and sanitation services (Jouravlev, 2005). Moreover, since the eighties the general trend of reforms in the sector have been carried out in a decentralisation process, in many cases at a very low jurisdictional level, i.e. townships.

The main arguments supporting this type of reform was based on the need to find a solution to local problems at local level among the local population in order to make the most of local initiatives and proximity to the users. This policy was strongly driven by the multilateral bank through loans to the sector and in many cases exploited by central and national governments to reduce the financial load that providing these services represented for their budgets (Fernández, 2009).

There are isolated cases in several countries where towns have managed to provide good drinking water and sanitation services. These situations are usually limited to large towns, with high income or political importance, which have managed to create autonomous companies managed by a relatively stable, professional and de-politicised directorate (Jouravlev, 2004). In general however, municipalisation has not led to providing more efficient services, and in many cases has given rise to small, inefficient, politicised service providers with little capability of recovering costs, technical difficulties in maintaining a quality service and financial restrictions preventing expansion of the service or improvement of environmental sustainability. The main problems identified with this based on experience in the region are described as follows:

Loss of economies of scale. The drinking water and sanitation industry are characterised by the existence of significant economies of scale. Regional and international experience alike prove that economies of scale exist in towns ranging from 100 thousand to nearly one million inhabitants, or with network volumes of 70 million cubic metres per year (Ferro & Lentini, 2010). With larger populations or volumes diseconomies of scale start to appear, although there are cases of constant economies of scale in even larger service providers (of up to more than 4 million inhabitants). In general economies of scale and constant economies of scale are predominant. With smaller values there are costs savings by amalgamating or consolidating smaller and medium size service providers, thus gaining in scale. It is important to point out there are reasons to believe that these estimates do not correctly capture all economies of scale present in providing these services, since the behaviour of many companies in the region and the rest of the world, particularly fully privatised services, confirm this statement. For example, the technology requirements (drainage versus latrines), service quality, access to water resources, cost recovery level or regulatory demands (treatment of provided water) usually entail higher costs for larger providers. More often than not these factors have an asymmetric influence that gives a misleading perception that the costs of smaller companies are lower and thus cause underestimates of economies of scale.

Incongruence between the industrial structure of the sector and the jurisdictional level exercising the function of regulation and control. Excessive fragmentation of the industrial structure makes regulatory activity more difficult. It is not viable to suppose that a universe of hundreds of suppliers can be effectively regulated or controlled.

Reducing the possibilities of cross-subsidies. By reducing the size of areas where services are provided, and perhaps by making them more uniform in socio-economic terms, decentralising limits the possibility of applying cross-subsidies (financing services with losses with other more profitable ones), making access to services by low-income population more difficult.

Providing services under political rather than technical criteria. Municipalisation involves a close relationship between the service provider and the local government, which in many cases has resulted in serious cases of politicisation of technical decisions, and misuse of public resources. Moreover, most municipalities lack the necessary resources to be able to effectively deal with the processes inherent to providing the services and depend on funds from other government levels for this.

Neglecting rural areas. Because of political dynamics at local level, municipal governments tend to assign higher priority to the needs of urban populations in detriment of rural communities.

Lack of incentives to protect water capture sources, control of water contamination, adaptation of the variability of supply sources and climate change make sense from a certain geographical scale. Because of the fact that political/administrative limits in municipalities are normally overlapping, and the natural limits of basins do not coincide, internalisation or the externalities linked to protecting water sources is not encouraged, in fact it becomes more complicated. Very often the basin that captures the water supplied to one town is actually located in the jurisdictional district of another one. Likewise, coordination between water extraction and served water discharge only makes sense at supra-municipal scales. In fact, in densely-populated basins, it is common for served water discharge areas from one town to be located upstream from water capture areas of other towns. Another important aspect is that small scale providers usually depend on a single supply source, making them more vulnerable to climatic variations.

As a result of the described problems, it has been recognised that more than being a problem of rigid options, it is important to structure balanced systems, adapted to national conditions, where legal and political attributes are assigned at adequate governance levels, in accordance with technical considerations, availability of resources, management capacity and above all objective criteria that permit exploiting economies of scale to the maximum and reducing transaction costs (Peña & Solanes, 2003). Carrying on with a strongly fragmented structure means that transferring the benefits of economies of scale to consumers, in terms of lower tariffs and better quality and more sustainable services, has to be revoked, as has been widely proved in international experience in this subject (Dupré & Lentini, 2000).

6.4.1 Implications of globalisation

Latin American and Caribbean countries have signed numerous agreements to protect foreign investment. The decisions by international courts of arbitration tend to restrict the power of governments to act in the benefit of public interests and local communities (Lee, 2007). This is obviously relevant in subjects related to regulating public utility services. At the same time, defending a number of consecrated rights of investors in this type of agreement actually compromises the regulatory functions of States. Three factors merge here (Jouravlev & Hantke-Domas, 2010):

Investment disputes generally deal with matters of public interest. Nevertheless, the mandate by arbitrators is to protect foreign investors. Consequently, the matters of public interest and the public utility services are removed from their mandates.

Arbitration investment courts can only assemble on request by the investors, which compromises their impartiality and due proceedings. For example, the decisions are made in secret procedures, without compulsory precedents or traditional appeals. Arbitrators are remunerated according to the cases they are assigned, where they can act as judges and lawyers at the same time, which leads to situations of potential conflicting interests. They also tend to adopt broad interpretations as to what expropriation is and violation of the due process.

There are a number of breaches between the jurisprudence of the international investment courts and the general principles of law in the nations, applicable to conflicts regarding economic regulation of drinking water and sanitation services.

These breaches are particularly wide in the following subjects (Jouravlev & Hantke-Domas, 2010):

Measures to face the impact of the economic crisis on public services. Arbitration jurisprudence, with a few exceptions, tends to ignore the criteria adopted by national courts, which in times of crisis are reluctant to increase tariffs, suspend payment, amend interest rates and suspend enforcement.

The common legal principles governing activity by private investors when providing public services, such as efficiency, due diligence, transparency and reasonable conduct, which are all but ignored by international courts of arbitration in spite of their undeniable relevance in the compared legislation.

Regulation according to public interest, which arbitration courts often assimilate with expropriation, whilst practically ignoring all the general principles of law in the nations concerning this matter.

Consequently, at the Seminar "International Investment Agreements, Sustainability of Infrastructure Investments and Regulatory and Contractual Measures" (Lima, Peru, 14–16 January 2009) the following items were proposed (Jouravlev & Hantke-Domas, 2010):

That countries should be unwilling to sign international investment agreements that do not contain public interest protection clauses and that do not claim the general principles of law of the nations in regulatory affairs.

That the rights of foreign investors are added to their correlative duties, such as efficiency, due diligence, good faith, transparency and respect for public interests in the host country and its legal rules.

To implement a reform of the investment arbitration system in order to correct the procedural dysfunctions inherent in the system, including lack of accountability, shortcomings in terms of publicity and participation, non-existence of arbitration jurisprudence unification and the lack of independence of arbitrators.

That the common regulatory principles that have been developed in the compared legislation, in terms of protecting public interests, public services and management of economic crises, are specifically included in the applicable law of investment arbitration. Investment protection agreements have led to a legal system through which international institutions have trampled over national law applying principles that have no bearing on the subject itself, nor coincide with the law that States apply under the same circumstances.

That before accepting international private investment in public services, countries should develop and implement their regulatory frameworks based on the fundamental principles of efficiency, good faith, equality before the law, due diligence and transparency.

That it should be made clear to investors that procedures before foreign investment promotion agencies do not exempt them from complying with all obligations before competent sectoral bodies, nor are they an excuse to ignore national legislation and regulations.

That investment agencies should not grant authorisation without specific agreement by the competent sectoral bodies in matters of public services, water resources and the environment.

6.5 CONCLUSIONS

Based on experience over the last two decades, we have learnt that there are certain fundamental principles in terms of providing drinking water and sanitation services, among which the following are mentioned (Jouravlev & Solanes, 2009):

Drinking water and sanitation services are hugely beneficial for public health, welfare, social equality, overcoming poverty, socio-economic development, environmental protection and political stability. Hence the priority that governments should give to this sector in their public policies is undeniable, particularly when referring to assigning budgets and construction of solid, stable sectoral institutions, even in times of crisis.

Efficiency reduces costs, which results in better opportunities for use. The most common inefficiencies are transfer prices, excessive debts, corruption, redundant labour, transaction costs, loss and scope of economies of scale, and capture by groups of interest. By artificially increasing costs, inefficiency damages equality. Hence, efficiency and equality are not mutually antagonising criteria - they are complementary to each other.

Decentralisation and municipalisation are not conducive to economies of scale, they cause transaction costs to increase and regulation and control activities to become more difficult, thus affecting efficiency and consequently equality. At the same time, evidence about economies of scale in this industry is conclusive. Hence economic regulation must promote and encourage amalgamation processes based on objective criteria.

In general terms there is nothing to justify a preference for public or private ownership. Consequently, a case by case assessment of the advantages and disadvantages should be carried out in accordance with the local conditions and established objectives.

Whether the company is public or mixed does not eliminate potential conflicts of interest in contractual matters, transfer prices, pay rises, redundant labour, or overcharging; the only thing that changes is the parties who benefit from them.

Governments should impose suitable regulations on public, private and mixed service providers, based on fair and reasonable profitability principles, useful and usable investment, good faith, due diligence, obligation of efficiency and transfer of efficiency gains to consumers.

In the case of public companies, the regulatory framework must be complemented with objective, personal responsibilities rather than institutional ones; including joint criminal, administrative and civil sanctions throughout the chain of command and management and directly responsible parties, in order to provide economically efficient services, i.e. seeking the sustainable alternative at the lowest cost for consumers; competitiveness for acquisition of raw materials and product production; and transparency when furnishing information. Criminal responsibility must be attributed to individuals, since it would be absurd for the State, through a mixed or public company, to hedge for acts by individuals.

Providing good quality services in a sustainable manner for the whole population requires suitable assessment of public policy decisions, according to rational economic, social and environmental guidelines, with emphasis placed on systematic application of the notion of economic efficiency; due consideration for the reality of national macroeconomics; and rigorous time-scaling of the economic, social and environmental goals.

Artificial guarantees and protection increase the risk of inefficiency and failure, since they provide unsustainable security and reduce the incentives to make efficient decisions. Hence, improving the decision-making process is indispensable. Countries should critically analyse their expansion alternatives (in terms of financing methods, technology, service methods, public guarantees, etc.) and structure them so that they do not become a burden for the economy and the citizens, eventually becoming a regressive factor working against growth.

If the national economy is unable to finance service provision, through salaries and taxes, private investors by themselves will contribute the additional economic resources as sunk costs, and therefore the services will never be sustainable. Privatisation cannot miraculously turn an unprofitable service into a profitable service.

In this sector, efficiency fundamentally depends on the regulatory framework and the institutional and structural conditions of the environs. Hence the "importance that governments give to equal distribution of the benefits of the reform is reflected in the professionalism the subject of regulation is tackled with" (Chisari *et al.* 1997).

6.6 REFERENCES

CEPAL (1992). La administración de los recursos hídricos en América Latina y el Caribe, LC/G.1694, Santiago de Chile.

CEPAL (1998a). Progresos realizados en la privatización de los servicios de utilidad pública relacionados con el agua: reseña por países de Sudamérica, LC/R.1697/Add.1, Santiago de Chile.

CEPAL (1998b). Progresos realizados en la privatización de los servicios de utilidad pública relacionados con el agua: reseña por países de México, América Central y el Caribe, LC/R.1697, Santiago de Chile.

CEPAL (1999). Tendencias actuales de la gestión del agua en América Latina y el Caribe (avances en la implementación de las recomendaciones contenidas en el capítulo 18 del Programa 21), LC/L.1180, Santiago de Chile.

CEPAL (2000). Equidad, desarrollo y ciudadanía, LC/G.2071/Rev.1-P, Santiago de Chile.

CEPAL (Comisión Económica para América Latina y el Caribe) (2005). Objetivos de Desarrollo del Milenio: una mirada desde América Latina y el Caribe, LC/G.2331, Santiago de Chile.

Chisari O., Estache A. and Romero C. (1997). Winners and Losers from Utility Privatization in Argentina: Lessons from a General Equilibrium Model. Banco Mundial, Washington DC.

Ducci J. (2007). Salida de operadores privados internacionales de agua en América Latina. Banco Interamericano de Desarrollo (BID), Washington DC.

Ducci J. and Krause M. (2012). Nota sobre regulación de empresas de servicios de agua y saneamiento de propiedad del Estado, mimeo. Banco Interamericano de Desarrollo (BID).

Dupré E. and Lentini E. (2000). Experiencia en América Latina. In: Privatización del sector sanitario chileno: análisis de un proceso inconcluso, S. Oxman and P. Oxer (comps.), Ediciones Cesoc, Santiago de Chile.

Estache A., Guasch J.-L. and Trujillo L. (2003). Price Caps, Efficiency Payoffs and Infrastructure Contract Renegotiation in Latin America. Banco Mundial, Washington DC.

Fernández D. (2009). Sustentabilidad financiera y responsabilidad social de los servicios de agua potable y saneamiento en América Latina. In: Contabilidad regulatoria, sustentabilidad financiera y gestión mancomunada: temas relevantes en servicios de agua y saneamiento, D. Fernández A. Jouravlev E. Lentini and Á. Yurquina (comps.), Comisión Económica para América Latina y el Caribe (CEPAL), LC/L.3098-P, Santiago de Chile.

Ferro G. and Lentini E. (2010). Economías de escala en los servicios de agua potable y alcantarillado, Comisión Económica para América Latina y el Caribe (CEPAL), LC/W.369, Santiago de Chile.

Ferro G., Lentini E. and Romero C. A. (2011). Eficiencia y su medición en prestadores de servicios de agua potable y alcantarillado, Comisión Económica para América Latina y el Caribe (CEPAL), LC/W.385, Santiago de Chile.

Foster V. (2001). Regulación del sector agua en la América Latina, Primer Encuentro de Entes Reguladores de las Américas (16 al 19 de octubre, Cartagena de Indias, Colombia).

Hantke-Domas M. (2011). Control de precios de transferencia en la industria de agua potable y alcantarillado, Comisión Económica para América Latina y el Caribe (CEPAL), LC/W.377, Santiago de Chile.

Hantke-Domas M. and Jouravlev A. (2011). Lineamientos de política pública para el sector de agua potable y saneamiento, Comisión Económica para América Latina y el Caribe (CEPAL), LC/W.400, Santiago de Chile.

Jouravlev A. (2003). Acceso a la información: una tarea pendiente para la regulación latinoamericana, Comisión Económica para América Latina y el Caribe (CEPAL), LC/L.1954-P, Santiago de Chile.

Jouravlev A. (2004). Los servicios de agua potable y saneamiento en el umbral del siglo XXI, Comisión Económica para América Latina y el Caribe (CEPAL), LC/L.2169-P, Santiago de Chile.

Jouravlev A. and Hantke-Domas M. (2010). Acuerdos internacionales de inversión, sustentabilidad de inversiones de infraestructura y medidas regulatorias y contractuales, Comisión Económica para América Latina y el Caribe (CEPAL), LC/W.325, Santiago de Chile.

Jouravlev A. and Solanes M. (2009). Las lecciones aprendidas en dos décadas de reformas sectoriales. *Recursos, Agua y Ambiente*, **3**(4).

Laffont J.-J. (1994). The new economics of regulation ten years after. *Econometrica*, **62**(3).

Lentini E. (2009). La contabilidad regulatoria de los servicios de agua potable y alcantarillado: la experiencia en el Área Metropolitana de Buenos Aires, Argentina. In: Contabilidad eregulatoria, sustentabilidad financiera y gestión mancomunada: temas relevantes en servicios de agua y saneamiento, D. Fernández A. Jouravlev E. Lentini and Á. Yurquina (comps.), Comisión Económica para América Latina y el Caribe (CEPAL), LC/ L.3098-P, Santiago de Chile.

Lentini E. and Ferro G. (2014). Políticas tarifarias y regulatorias en el marco de los Objetivos de Desarrollo del Milenio y el derecho humano al agua y al saneamiento, Comisión Económica para América Latina y el Caribe (CEPAL), LC/L.3790, Santiago de Chile.

Peña H. and Solanes M. (2003). La Gobernabilidad efectiva del agua en las Américas, un tema crítico, III Foro Mundial del Agua (16 al 23 de marzo de 2003, Kioto, Japón).

Rodríguez J. (2002). Contabilidad regulatoria. Aplicación en el sector sanitario chileno, Superintendencia de Servicios Sanitarios (SISS), Santiago de Chile.

Solanes M. (1999). Servicios públicos y regulación. Consecuencias legales de las fallas de mercado, Comisión Económica para América Latina y el Caribe (CEPAL), LC/ L.1252-P, Santiago de Chile.

Solanes M. (2002). América Latina: Sin regulación ni competencia? Impactos sobre gobernabilidad del agua y sus servicios, mimeo, Comisión Económica para América Latina y el Caribe (CEPAL), Santiago de Chile.

Solanes M. (2007). Formulación de nuevos marcos regulatorios para los servicios de agua potable y saneamiento, Carta Circular de la Red de Cooperación en la Gestión Integral de Recursos Hídricos para el Desarrollo Sustentable en América Latina y el Caribe, N° 26, Comisión Económica para América Latina y el Caribe (CEPAL), Santiago de Chile.

Jean-François V. (2010). Experiencias relevantes de marcos institucionales y contratos en agua potable y alcantarillado, Comisión Económica para América Latina y el Caribe (CEPAL), LC/W.341, Santiago de Chile.

Chapter 7

The German experience with a self-organised water sector – key factors for an alternative to regulation

Wolf Merkel and Nicole Annett Müller
IWW Water Centre, Moritzstraße 26, 45476 Mülheim an der Ruhr, Germany

7.1 INTRODUCTION

In Germany, the water supply and wastewater removal sector can be characterised with high performance, quality orientation and reliability for the customer. Further, the technical conditions of the systems as well as environmental and hygienic standards are at a high level (see Libbe, 2014: 1), the drinking water quality and the reliability of supply (quantity, location, time) are excellent (see Umweltbundesamt, 2010: 72; Bundesministerium für Gesundheit, 2011: 2). The greatest challenge is to maintain these quality levels with limited resources. Therefore, discussions among the sectors' stakeholders focus more and more on economic performance. The aim of this paper is to present to what extent the German challenges can be met under a sectoral structure that is primarily based on self-organisation.

7.2 GENERAL CHARACTERISTICS

Compared to many other European countries, Germany is fortunate to have plenty of water resources. Besides many natural lakes, rivers and well-fed groundwater systems in large areas, Germany has numerous artificial lakes and reservoirs. Only 2.7% of the available water resources are used for public water supply, which indicates extensive water reserves (see Umweltbundesamt, 2010: 14–18; Statistisches Bundesamt, 2011: 20–25). Further, the impacts of climate change are likely to be moderate. Nonetheless, the long-term planning and investment periods in the sector make forecasts of and adaptations to the consequences of climate change necessary (see DVGW, 2009: 2). Besides precipitation during all seasons, the German climate is characterised by moderate temperatures and frequent weather change (see Statistisches Bundesamt, 2011: 20). The German Weather Service generally predicts sufficient future rainfall, whilst temperature forecasts based on different climate scenarios consistently indicate increasing temperatures (see Deutscher Wetterdienst, 2014a, 2014b). It can be summarised that "across Germany, annual average temperatures will go up, resulting in warmer and drier summers and milder and wetter winters" (DVGW, 2009: 2).

Apart from climate change, Germany will also be affected by demographical change. Whilst the population was around 82 million in 2008, future forecasts show an estimated decrease of around 4.6 million people by 2030 (see Statistische Ämter des Bundes und der Länder, 2011: 21). Especially for sectors with rigid assets and capacities like the water industry, strongly declining consumer numbers are alarming. Moreover, the ageing population becomes more and more challenging for wastewater treatment because of drug residues in the wastewater (see BDEW *et al.* 2011: 41).

7.3 LEGAL FRAMEWORK

The primary source of law for water supply and wastewater removal in Germany is the Water Resources Act (Wasserhaushaltsgesetz – WHG) of 2009. This law makes requirements concerning the management of all water resources: surface water, groundwater and marine water. Furthermore, the WHG includes regulations concerning flood protection, water body development, water supervision and fines. In addition to these themes, the law also deals with specific regulations concerning water supply and wastewater removal, of which the most important issues are presented in the following (see WHG, 2009).

Sewage shall be removed to a level that ensures that public welfare is not compromised. Further, the legal entities of public law, which are required under state law, are responsible for the wastewater removal. However, the liable entities are allowed to transfer the wastewater obligations to third parties. Anyone who operates a sewage system is required to maintain its state, its ability to function and its conservation. Moreover, the service provider has to monitor its operation and the type and quantity of sewage content himself. The entity has the obligation to record and store the relevant information and upon request to provide that information to the competent authority (see WHG, 2009: § 54–61). Since 2009, water supply has *officially* been a service of general interest (Daseinsvorsorge), which emphasises its great importance and essentiality. The whole population depends on an adequate water supply. For this reason, the water suppliers are forced by the requirements of the WHG to manage the resource water carefully and to inform the end-consumer of water saving opportunities. Further, the water demand shall be covered primarily by local, close water resources, if the effort to do so is reasonable and acceptable (see WHG, 2009: § 50). Thereby no region shall be affected disproportionately in applying area-wide water conservation (see Lotze & Reinhardt, 2009: 3277).

The hygienic requirements for the provided drinking water are regulated by the German drinking water regulation (Trinkwasserverordnung) based on the EU Drinking Water Directive (see The Council of the European Union, 1998). The water quality regulations are controlled by the local health authority and reported to the German central government and to the EU commission. In case of inadequate drinking water quality, the local health authority initiates further action. Optionally, fines may be imposed or even the interruption of supply may be arranged (see Trinkwasserverordnung, 2001: § 9, 18, 24). Analogously, all requirements concerning wastewater are embedded in the wastewater regulations (see Abwasserverordnung, 1997), based on the European Wastewater Directive (see The Council of the European Communities, 1991).

Besides environmental and health regulations the water services also underlay structural requirements. In § 28 of the Basic Law (Grundgesetz) it is defined that the municipalities have the right to regulate all affairs of the local community on their own responsibility, however, they have to take account of the current legislation (see GG, 1949: § 28). Nonetheless, the right to self-govern the water services as part of the municipal public duty does not mean that it has to be fulfilled directly by the municipalities. Unless state law provisions prevent this possibility, the municipalities are allowed to transfer tasks to third party, private entities or make use of cross-municipality solutions (see Ewers *et al.* 2001: 17). Further, regulations concerning organisational issues are made at the state level by the State Water Laws (Landeswassergesetze).

The leading organisations in the field of water and wastewater are the Federal Association of Energy and Water (Bundesverband der Energie- und Wasserwirtschaft – BDEW), the German Association for Gas and Water for the Drinking Water sector (Deutscher Verein des Gas- und Wasserfaches – DVGW) and the German Association for Water, Wastewater and Waste (Deutsche Vereinigung für Wasserwirtschaft, Abwasser und Abfall – DWA) for the wastewater sector. Generally, the BDEW focuses on political and economic issues, whereas the DVGW and DWA prioritise the technical and organisational challenges of the sectors. Moreover, DVGW and DWA are issuing bodies for the relevant technical guidelines, based on the contributions of sector experts from the industry and confirmed in an open participatory process.

A general regulatory authority does not exist in the German water industry, so it is in fact a kind of technical self-regulation. "The legal framework defines basic requirements for the quality, safety, sustainability, and economic efficiency of water services, but it is the water sector itself which fills the legal framework with life through the definition of technical rules and standards" (Petry & Castell-Exner, 2012: 18). Nevertheless, economic issues like water pricing are regulated ex post to avoid the suspicion that the monopoly has been exploited. Depending on the organisational form, various regulatory authorities can take action, which is considered in more detail in the section *Economic Challenges* (see Dierkes & Hamann, 2009: 33).

7.4 DIMENSION AND STRUCTURE

In cooperation with other associations and stakeholders, the Federal Association of Energy and Water (BDEW) frequently publishes the *Profile of the German Water Sector*, which includes many statistics and further information regarding the water sector. This section presents the most relevant figures and numbers characterising the dimension and structure of the German water market. In 2011 around 4968 million m³ of water was produced. A total of 61.1% of the used water was groundwater, 30.5% was surface water and only 8.4% was spring water. Therefore, the groundwater recharge is a relevant issue in Germany. In comparison to 1990 the amount of total treated water indicates a decrease of around 1800 million m³ or 27% (see BDEW, 2013: 2).

In total, the drinking water supply in Germany is organised by 6211 companies. These companies serve 99% of the German population via an approximately 530,000 km-long network (see BDEW *et al.* 2011: 34, 52). Besides the challenge of a declining population, the water consumption per capita of the remaining population is also decreasing, even though previous forecasts assumed an increasing per capita consumption. This development is presented in Figure 7.1. The current water consumption of around 122 litres per person per day leads in many places to underutilisation of networks, which might be associated with hygiene challenges.

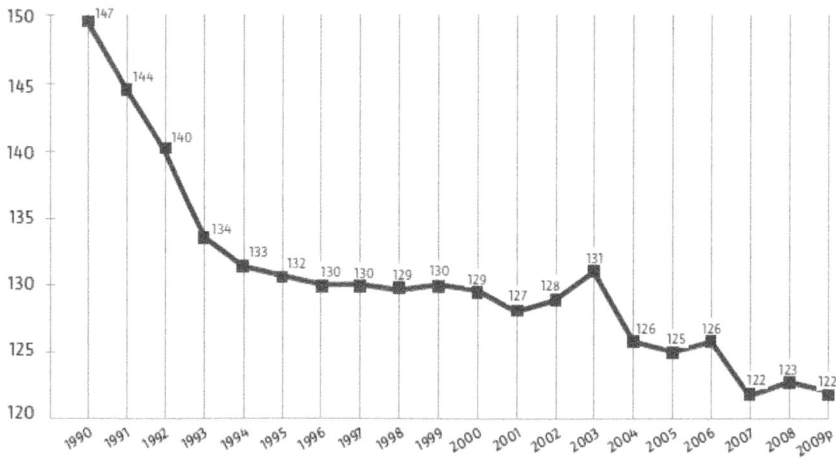

Source: BDEW Water Statistics, related to households and small trades. p = provisional

Figure 7.1 Development of the per-capita water consumption in Germany (in litres per person per day) (BDEW *et al.* 2011: 39).

The reliability of supply can be described as high. "Long, frequent service interruptions of water supply are unknown in Germany. This is due to the high technical standards and the excellent condition of plants and networks in comparison with other European countries. German water supply utilities have by far the lowest water losses" (BDEW *et al.* 2011: 48). Average water losses in Germany amount to approximately 6.5% and the rate of main failures has further decreased during the last decade. Whilst in the period of 1997–2004 11.7 damages occurred per 100 km network length, the number of main failures decreased to 9.9 per 100 km between 2005 and 2009. Nevertheless, benchmarking projects in different German states identify varying renewal rates from 0.4 to 1.2% (see BDEW *et al.* 2011: 56f.). The renewal rates are very important in terms of sustainability. It is great to achieve good technical and qualitative standards, but the sector has also to ensure the future viability of the network. This fact similarly applies to the sewage network.

The current sewage connection rate amounts to 96.1% in Germany. Ninety-five per cent of the population is connected to wastewater treatment plants in accordance with highest technical EU standards. The wastewater piping system is estimated to amount to 187,264 km, the storm water piping system to 114,373 km and the combined water system is around 239,086 km long. Considering the age structure of the German sewer network, it is striking that approximately 70% of the pipes are younger than 50 years. Nevertheless, parts of the main system are much older. In order to ensure a sustainable maintenance, continuous investments are necessary (see BDEW *et al.* 2011: 52ff., 68).

On account of the legal framework in Germany, the municipalities have the option to fulfil tasks of general interest on their own or in collaboration with third parties. Among collaboration models, different configurations are possible. The common management models can be divided into models under private and public law.

The public forms of organisation are mainly:

- Ancillary municipal utilities (Regiebetrieb)
- Owner-operated municipal utilities (Eigenbetrieb)
- Institution under public law (Anstalt des öffentlichen Rechts)
- Special purpose association (Zweckverband)
- Water and soil association (Boden- und Wasserverband)

The Regiebetrieb is fully embedded in the municipality. It is legally and organisationally dependent and does not have separate accounting, so that surpluses are allocated to the General Fund. The Eigenbetrieb is also legally dependent, but organisationally and financially independent from the municipality. Therefore, losses and profits are earmarked. The organisation of water supply as an institution under public law offers the strongest independence from the community. It is organisationally, financially and legally independent. Intermunicipal cooperation is possible via different association forms (Zweckverband/Boden- und Wasserverband) (see Cronauge, 2003: 30ff., 114ff., 225ff.; Dierkes & Hamann, 2009: 169ff.).

Moreover, different private models exist in the German water market, from mixed public-private companies to autonomous private companies. Especially, concessions play a major role in public-private-partnerships and are further discussed in the section on *Competition*. In the following the ownership structure of the drinking water and the wastewater sector as well as its development shall be presented.

As mentioned above, the total drinking water supply in Germany is organised by 6211 companies. BDEW *et al.* (2011) have evaluated detailed characteristics of around 1218 companies, which account for 75% of the total water production, ensuring the representativeness of the results concerning the structure of the German water market. This analysis shows that in 1993 aggregate water output was divided in almost equal shares between public and private suppliers, although the number of public water supply utilities was significant higher (see Figure 7.2). This confirms the assumption that private parties primarily operate in densely-populated areas. Comparing the numbers of 1993 to the results of 2008, there is a tendency towards private organisational models, mostly in parts or entirely owned by the public entities.

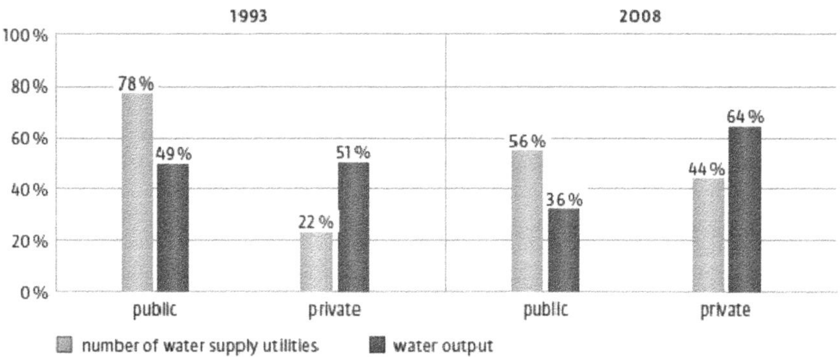

Figure 7.2 Development of the types of enterprise in the public water supply (under public/private law) (BDEW *et al.* 2011: 34).

The detailed ownership structure in the German water market is represented in Figure 7.3. Under consideration of the water output, mixed public-private companies dominate the market with 26%, followed by special purpose associations (17%) and other private-law utilities (16%). With only 1% of the total water output, ancillary municipal utilities play almost no role in the drinking water sector.

Besides the ownership structure, the size structure in the German water market is also of great interest. Figure 7.4 shows that less than 4% of water utilities provide 60% of the total water output in Germany. Further, it indicates that around 70% of the water utilities are small companies with a water output of less than 0.5 million m³ per year. The total number of 6211 water utilities combined with the results of Figure 7.4 underline the fragmented structure of the drinking water market as well as its diversity. The largest German (end-user) water supply companies are Gelsenwasser, Berliner Wasserbetriebe, Stadtwerke München and Hamburg Wasser.

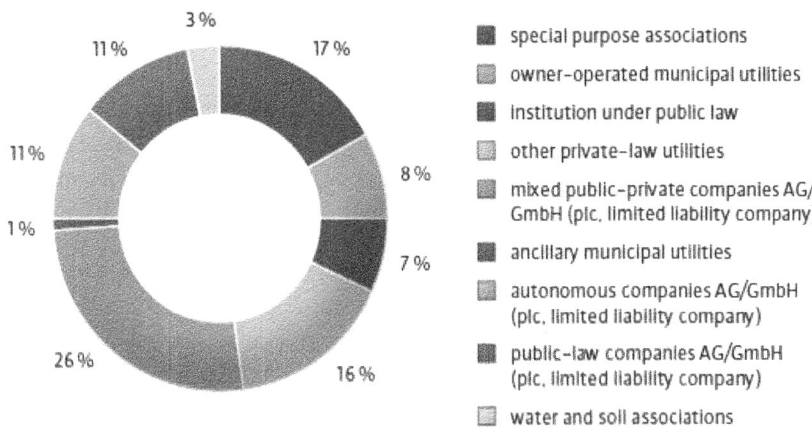

Figure 7.3 Types of enterprise in the public water supply 2008 (shares related to water output) (BDEW *et al.* 2011: 35).

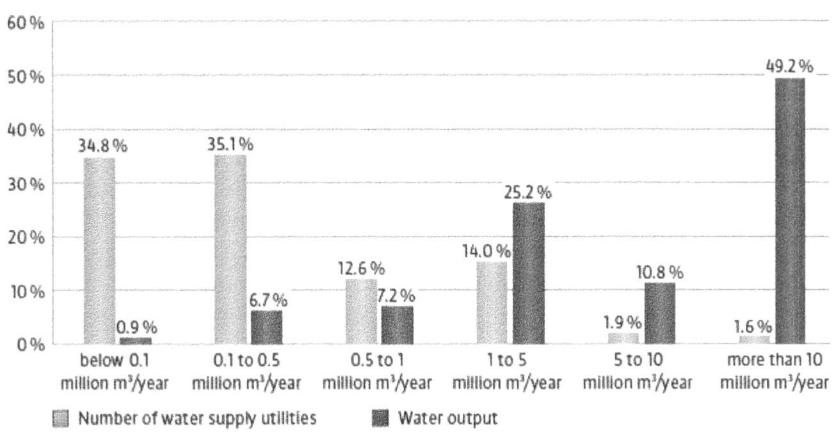

Figure 7.4 Size structure of water supply utilities in Germany 2007 (shares as per cent) (BDEW *et al.* 2011: 36).

In contrast to the drinking water sector, almost every wastewater utility acts under public law. Private participation plays only a subordinate role in the sewage sector. The most common organisational form is the owner-operated municipal utility (37%), followed by different intermunicipal associations (28%) and institutions under public law (13%). The size structure of the German wastewater market is similar to the one in the drinking water sector. Although the market can be described as fragmentised, a few large providers take care of the wastewater services in metropolitan areas (see BDEW *et al.* 2011: 36ff.).

7.5 COMPETITION

The legal framework fosters the utilisation of local water resources, thus, favouring a fragmented water supply in Germany. Moreover, due to the natural monopoly there is a lack of competition in the German water market. For physical and economic reasons, usually only one local water utility can prevail. The mounting of a second or further system in the field of water (and sewerage) is very expensive and not considered economically feasible (see e.g. Dierkes & Hamann, 2009: 17). Further, an adequate legal framework for common carriage does not exist. However, even if the responsibilities could be clarified under an appropriate law, common carriage approaches will fail in most cases due to cost aspects. On the one hand, technical measures have to meet the requirements of a chemically and hygienically safe water mixture. On the other hand, the utility that wins the resource close to the supply area will always have a competitive advantage against utilities, which have to rely on water from distant regions, due to lower transport costs (see Mankel, 2002: 42f.).

The municipalities' opportunity to tender the water services or parts of it can help to force more competition for the market (see European Commission, 1999: 40). Companies are bidding for the right to fulfil the service (or parts of it) in a given period of time. Moreover, in most cases further requirements are determined in the contract like, for example, the level of customer service or environmental protection (see Oelmann, 2003: 5). The municipality can choose the company with the best offer, which generally increases the pressure on the companies to compete in questions of expertise, efficiency and prices. Therefore, Demsetz (1968) already speaks about the "best price-quality package", which has to be considered. In Germany, concessions are the dominating model for the involvement of third private parties. These concession contracts affect typically the task fulfilment and not the task responsibility (see Dierkes & Hamann, 2009: 143ff.). Whilst a recent European directive on the award of concessions fills the absence of clear requirements by an "[...] adequate, balanced and flexible legal framework [...]", concessions in the water sector are excluded from this requirements since they "[...] are often subject to specific and complex arrangements which require a particular consideration given the importance of water as a public good of fundamental value to all Union citizens" (The European

Parliament and the Council, 2014: 1, 40). In addition to concessions, operator models, management models and cooperation models are applied in the German water and sewage sector (see Dierkes & Hamann, 2009: 143ff.).

This process creates a kind of limited competition *for* the market. Nonetheless, the contracts are typically long-term, so that this form of competition is mainly restricted to the time of the tender. Further, the advantages of a competitive concession can be missed, if the number of bidders is too low or (in the worst case) if there is no other competitor (see e.g. Garcia *et al.* 2005: 173, 180). The water and sanitation business is very specific, which naturally limits the number of potential bidders. Consequently, mostly the same players are active in a country's water market.

7.6 ROLE OF BENCHMARK ACTIVITIES

Besides the competition *for* the market, a competition *via comparison* is of particular interest in the German water market. Different benchmarking projects, efficiency analysis as well as price comparisons are omnipresent in the German water industry. The publication of current data and evaluation results informs the end-user about opportunities, simultaneously increasing the pressure on the providers to operate more efficiently. Customers' education and sensibility for water and wastewater related questions is necessary to foster the end-users' influence on the performances.

The concept of benchmarking has gained significant attention, importance and acceptance in the water sector during recent decades (see Carvalho *et al.* 2012). In the context of this development several definitions and methodologies have been raised. It can be commonly stated that "benchmarking is a tool for performance improvement through systematic search and adaptation of leading practices" (Cabrera *et al.* 2011: 2). A fundamental part of the concept is learning from others as well as having a closer look at internal processes to analyse which improvement actions are needed and how they can be implemented. As part of a benchmarking process, which is usually based on performance indicators, preparation and data acquisition, performance assessment via comparison as well as performance improvement are generally conducted. Whereas the first two stages are in most cases supported by the benchmarking organiser or respective regulator, the performance improvement mostly remains a task of the utility (see Cabrera *et al.* 2011: 2, 28; 95–96). The success of these improvement measures can ideally be proven during the next benchmarking event.

According to its scope, benchmarking can be divided into metric benchmarking approaches (total approaches) and process benchmarking approaches. Metric benchmarking is a comparison of comprehensive performance indicators of different but at best similar utilities. It helps to provide comprehensive information, to identify the best and the weakest performance within a utility sample and is therefore a comparison tool, which is often used by regulators or public organisers to enhance performance pressure and competition respectively (see Cabrera, 2008: 5). In contrast, process benchmarking is less general in nature and focuses on specific processes within production. Thus, it offers the possibility to identify specific stages within the production that need further attention and assessment (see Berg, 2010: 7).

In Germany, there are many regional benchmarking projects. In most German states (Bundesländer) water utilities have an opportunity to participate in non-obligatory projects (metric as well as process benchmarking). Eleven out of sixteen states also provide a public report with general performance including efficiency results, which is typically anonymous. It has to be noted that three of the provinces not providing a public project report are city states, so with publication of results anonymity would be lost. Generally, the benchmarking projects are carried out periodically, so the utilities can see performance changes over several years (see BDEW, 2012).

Figure 7.5 gives an indication of the actual participation rate for each state. Even if benchmarks are non-obligatory in Germany, many utilities make use of this opportunity. Nonetheless, it has to be stated that in some federal states (Bundesländer), the participation rates are obviously far from average, especially when the number of utilities participating in benchmarking exercises is considered.

The benchmarking largely sticks to different performance indicators in the fields of reliability, quality, customer service, sustainability and economy. Benchmarking of water supply services is based on the International Water Association (IWA) standard for performance indicators (see Alegre *et al.* 2000), issued in a German adaptation by Hirner and Merkel (2005). To enable, sooner or later, nationwide benchmarking, a unification of these indicators for all regional benchmarking is discussed (see Otillinger, 2011: 26). Besides regional and national benchmarking efforts, trans-national comparisons can help to share experiences and learn from the best. As an example of trans-national benchmarking, that between Austria and the German state of Bavaria can be highlighted (see Theuretzbacher *et al.* 2005).

Otillinger (2011) emphasises that the German benchmarking activities show a positive, growing and dynamic development. Since there are many ongoing benchmarking projects in Germany, the results of two selected projects shall be representatively summarised. Hein and Merkel (2010) point out that the process benchmarking in 15 drinking water plants in Germany helps participants to achieve transparency regarding process-related costs. Knowledge of their own cost ratio and cost types compared to the ones related to other treatment plants can be a starting point for future strategic decisions and increased process efficiency. The consideration of specific processes allows deeper insights and is therefore a good addition to metric, comprehensive benchmarks. Similar water supply conditions and cultural, legal and socio-economic characteristics have made cross-national benchmarking between Bavaria and Austria possible. Theuretzbacher *et al.* (2005) describe this trans-national benchmarking as very successful. In addition, the fact that results show performances which are comparable or even above international level, means it can be taken as a good example to further intensify international benchmarking cooperation.

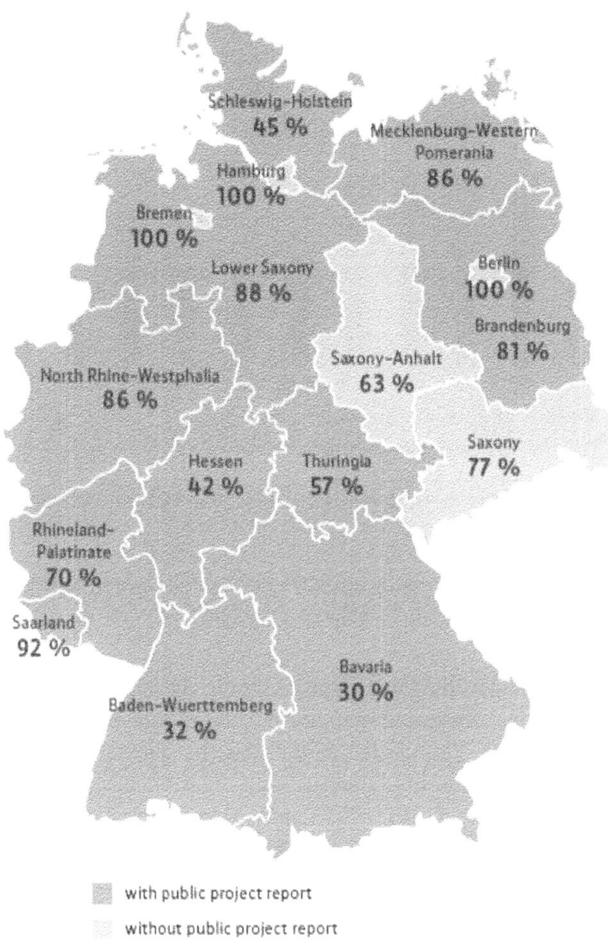

Figure 7.5 Participation rate of water utilities (measured by drinking water provided) (BDEW, 2012: 13).

7.7 ECONOMIC CHALLENGES

To ensure a sustainable water supply and sewerage service for the future, further investments are indispensable. This also applies to Germany, if the current service level is to be maintained. The main target is to make continuous investments to avoid unplanned high expenditures and related price increases. The investments of the water supply and wastewater removal industry have amounted to approximately 110 billion euros since the German reunification in 1990 (see BDEW *et al.* 2011: 75). The development of capital expenditure in both sectors is shown in the following figures.

During recent years the capital expenditure level in the water supply sector has been relatively constant at about 2 billion millions euros. It is striking that the bulk of investment flows in the pipeline network, whereas the investments in abstraction and treatment are decreasing (see Figure 7.6).

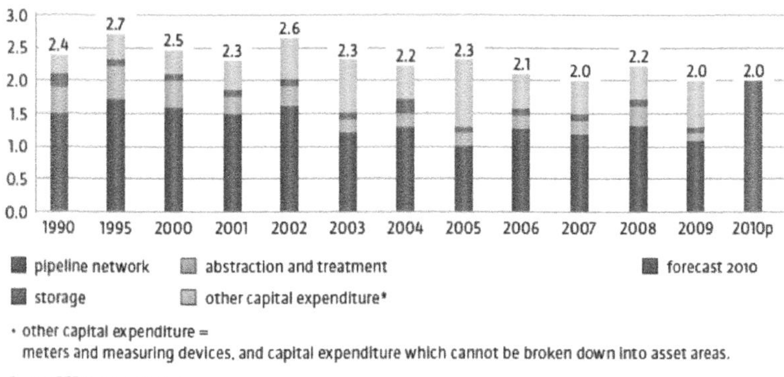

Figure 7.6 Development of capital expenditure in public water supply from 1990 to 2010 (according to asset areas, in billion euros) (BDEW *et al.* 2011: 76).

The development in the wastewater sector is characterised by a less homogeneous trend. As illustrated in Figure 7.7, the capital expenditures decreased significantly after 2000. This is due to the fact that capital investments related to the implementation of the EC Directive on Urban Wastewater Treatment were phased-out (see BDEW *et al.* 2011: 76). Comparing Figure 7.6 and Figure 7.7 it is noticeable that capital expenditure in the wastewater sector is more than twice as large as that in the water supply sector.

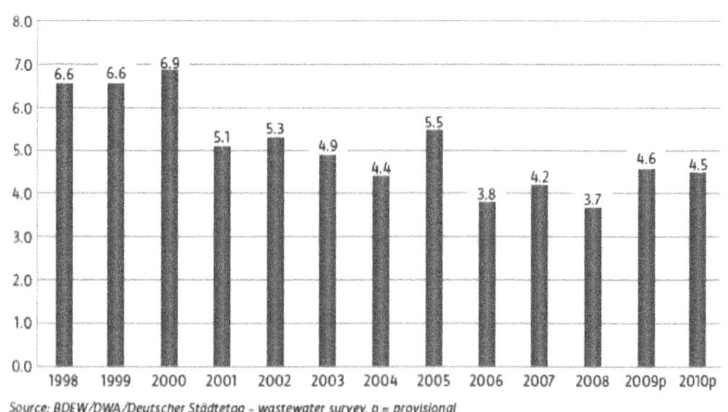

Figure 7.7 Development of capital expenditure in public wastewater supply from 1998 to 2010 (in billion euros) (BDEW *et al.* 2011: 77).

Subsidies play a minor or almost no role in the German water supply. The more important issue is a cost-covering price structure. Depending on the company's legal form, the water pricing is influenced by different frameworks. Whilst requirements for charges of companies under public law are made in the Municipal Charges Acts (Kommunalabgabengesetze – KAG) of the different states, water prices of companies under private law are not subject to specific regulations (see Reif, 2002b: 52). "However, according to the rulings of the German Federal Supreme Court, the principles applied to the calculation of charges are to be applied in the same way to the calculation of prices" (BDEW *et al.* 2011: 23). According to BDEW *et al.* 2011, the main obligations and principles are:

- Principle of equivalence (proportionality)
- Principle of cost recovery
- Prohibition of cost overrun
- Principle of equality or equal treatment
- Economic principles

Whilst the first obligations are self-explanatory, the economic principles may include the principle of preservation of net real-asset values or the principle of real capital preservation.

In the following Figure the supervision and control of prices and charges is illustrated. It has to be emphasised, that the participation of private companies does not automatically lead to the collection of prices (see Reif, 2002b: 53). The decisive factor is the legal form of the charging company. As presented in Figure 7.8, companies under public law can choose between charges and prices, whereas companies under private law are bound to prices.

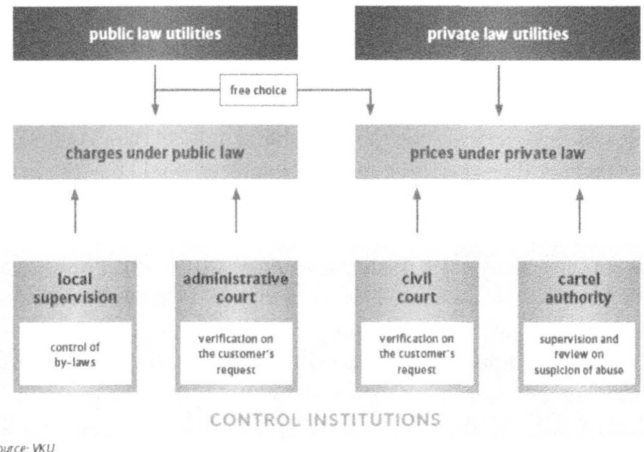

Figure 7.8 Control of prices and charges (BDEW *et al.* 2011: 24).

In the German drinking water supply sector a two-part tariff model dominates. Whilst the companies have to face extremely high fixed costs of about 80%, the fixed share of the tariffs is, at about 10–20%, very small (see Umweltbundesamt, 1998: 51; Merkel, 2009: 78; VKU, 2011: 2). In times of declining water consumption, this difference between the companies' cost and revenue structures leads to a cost coverage gap. This means that German "water utilities sell less water to fewer people at increasing costs for infrastructure maintenance and renewal" (Petry & Castell-Exner, 2012: 19). Figure 7.9 illustrates the effects of decreasing water deliveries on total and specific costs.

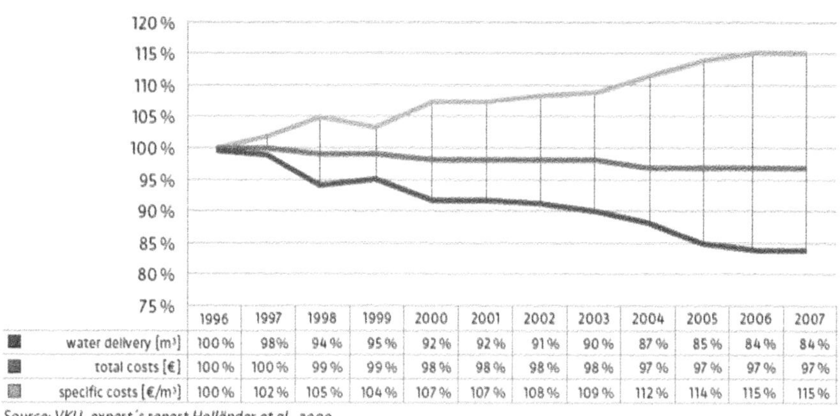

	1996	1997	1998	1999	2000	2001	2002	2003	2004	2005	2006	2007
water delivery [m³]	100%	98%	94%	95%	92%	92%	91%	90%	87%	85%	84%	84%
total costs [€]	100%	100%	99%	99%	98%	98%	98%	98%	97%	97%	97%	97%
specific costs [€/m³]	100%	102%	105%	104%	107%	107%	108%	109%	112%	114%	115%	115%

Source: VKU–expert´s report Holländer et al., 2009

Figure 7.9 Effects of decreasing water deliveries on total and specific costs (Relative evolution over time) (BDEW *et al.* 2011: 43).

It is striking that the specific costs significantly increase with declining water consumption. Therefore, there is currently a hot debate on adjustments of tariff designs in Germany. These adjustments are necessary on the one hand to ensure a sustainable cost recovery and on the other hand to meet the principle of equality or equal treatment. Otherwise, high usage consumers would pay the fixed costs for low usage consumers (see e.g. Oelmann & Haneke, 2008).

Currently the water prices strongly differ over the country; differences of up to a factor 6 were observed (see Hirschhausen *et al.* 2010: 76). This situation raises questions about price fairness among the population. Although price comparisons are not evident, because different local circumstances require different cost structures, the cartel authority of the state of Hessen has imposed requirements for one water supplier to lower its water prices by 2007. This judgment was affirmed by the Federal Court in 2010 (see Daiber, 2010: 226). To avoid the decision, the affected company, which was under private law, has now decided to operate under public law in the future. This case also offers a basis for discussion among the population as well as in the water sector. Especially, the issue of the legally-accepted tariff with regard to price calculation has not been clearly resolved, in spite of current initiatives from the water associations for a price calculation guideline (see Reif, 2002a: 19; BDEW, 2010; BDEW & VKU, 2012).

The prices for drinking water include a reduced value added tax of 7% (normal 19%) and – depending on state law – a water abstraction levy. These water abstraction levies vary between 0 and 31 Cent per m³ of abstracted water. Furthermore, the revenues from the water abstraction levies do not have a tied purpose in every state (see BDEW *et al.* 2011: 27f.). Between 2005 and 2013 the volumetric charges increased by approximately 8%, whereas the basic charges increased during the same period by around 19% (see Statistisches Bundesamt, 2014: 1). In 2010 the average water price for households amounted to 1.91 Euro per m³, which already includes the share of the fixed price component (see BDEW, 2011: 5). Considering the current daily water consumption of around 122 litres, this implies average annual water costs of around 85 euros per person.[1]

The taxation of wastewater services is more differentiated. "Public wastewater disposal utilities as sovereign undertakings are exempt from corporate income and turnover tax. If a utility responsible for wastewater disposal uses a private third party to discharge this obligation, the latter is subject to the full turnover tax rate with the possibility of input tax deduction" (BDEW *et al.* 2011: 29). Similar to the water abstraction levy, the utilities have to pay a wastewater tax to the respective state and pass these additional costs directly to the end-consumer. In 2008 the revenues of the wastewater tax amounted to 254.05 million euros (see EUWID, 2011: 32). The average price per m³ of wastewater (according to the freshwater scale) was around 2.28 euros in 2005. Moreover, many utilities divide the according scale into storm water and sewage. In these cases, the average prices evaluated in 2005 amounted to 2.05 euros (sewage) and 0.88 euros (storm water) (see BGW/DWA, 2008: 2f.). Thus in Germany the average wastewater charges are generally higher than the charges in the drinking water sector.

Although water fees and their diverging ex post regulation are moving more and more into the focus of public debate and cartel authorities (see Daiber, 2010), it remains to be stated that the prices are generally affordable and the water costs do not play a decisive role in total household budget.

7.8 LESSONS LEARNT FROM THE GERMAN CASE – GOOD WATER SERVICES WITHOUT A REGULATOR?

The performance of municipal water services is to be evaluated in terms of the *service* they offer to their citizens. In general, the criteria for good water services are i) quality of water (drinking water and wastewater), ii) quality of service (reliability of supply/disposal), iii) sustainability of services in financial and

---

[1]Estimation based on the calculation ((122 × 365)/1000) × 1.91 = 85.05.

technical terms and iv) the affordability in terms of adequate costs in relation to the service delivered.

Water/wastewater quality is determined by European and national legislation, and is regulated and controlled by the German government. With regard to the quality of service and the technical sustainability, the key success factor of the German water sector is the self-organised process of setting technical guidelines. In these technical guidelines, issued predominantly by DVGW and DWA, the water sector sets itself standards for good services in technical and organisational terms. In the example of water supply, the German Drinking Water Directive refers to these technical guidelines in the general clause to work *according to existing technical standards*. The standards specify relevant issues of good water services: continuity, minimum pressure, choice of processes, requirements for materials, prohibition of improper installations, testing procedures etc. In case of a malfunctioning of service, the water and/or wastewater utility – respectively the responsible engineering companies, suppliers, contractors etc. – have to prove that they acted according to the standards, or need to explain deviations under the specific circumstances. This system of self-regulation works quite effectively, is widely accepted among utilities and the government, and directs the optimum (self-) regulatory power to areas where it is needed most. The main drawbacks are that this democratic process of guideline-setting usually takes its time, and that the criteria of cost-effectiveness have to be enforced within the process.

The effective implementation of these guidelines in utilities of all sizes is of course an issue, and there are doubts as to whether particularly small water and/ or wastewater utilities have the manpower, knowledge and financial means to implement proper management and operation according to the guidelines. The implementation is enforced by regular governmental inspections, which is a quite an effective process, but, however, not unlikely to fail under certain circumstances.

There is also a system of economic supervision in place for all water services, implemented at the regional government level, and a control system established at the level of the states' government (Landesrechnungshof). Thus, the supervision and control system should secure the technical and the economic sustainability of municipal water services. On average, this clearly is the case, but there are cases known where cost-coverage and renewal levels have been insufficient.

As pointed out before, there is a wide spread of prices/tariffs all over the country, which in itself is not a problem. There are areas, where the provision of drinking water or the adequate disposal of wastewater require more sophisticated infrastructure and are more expensive than in other regions. From the citizens' point of view, these circumstances are not always obvious, with the result that public debates arise regularly. Tariff or price control for water services – as for the control of other public services – are also accepted governmental responsibilities. Citizens have the right to request price control concerning tariffs, and in recent years, governmental institutions in several federal states have taken the initiative for a formal price check, using data from comparable utilities for the evaluation.

So even without direct economic regulation, there are supervising functions installed in governmental institutions, which effectively protect citizens from the utilities' abuse of their monopolistic position.

7.9 CONCLUSION

The described challenges, organisational structure and technical level of the German water sector cause different views and opinions on its self-organisation. Boscheck (2002) highlights that proponents of the self-regulatory model see the good status quo of the systems under an individual public responsibility reflecting the tradition of municipal subsidiarity, whilst opponents of the model criticise that the high fragmentation level being associated with the model causes particular inefficiencies in operations and investments.

Measuring the success of a certain system or an organisational design is not trivial. On the one hand, a direct comparison to another alternative water sectoral structure facing exactly the same challenges and characteristics is not possible. On the other hand, appropriate performance indicators have to be identified, which allow a conclusion regarding the success. In general, it can be stated that a water sector is working, if technological requirements (connection to the network, reliability of supply etc.) and quality standards are met *now* and if this form of structural framework ensures that these targets will also be met in the *future*, at both times under the constraint of price affordability.

As presented in this paper, the German water sector is currently characterised by good technical conditions and excellent drinking water quality. The prices are affordable even though they are not equal over Germany due to regional circumstances. Furthermore, the reputation of the German water utilities is very good as around 91% of the customers are satisfied with the service of their local water supplier and about 77% are satisfied with their sewage company. Nonetheless, it has to be stated that many Germans (around 65%) are not aware of their actual water consumption or their specific annual water costs (see BDEW *et al.* 2011: 62, 65). On the one hand, this is an indication of the drinking water's affordability as the consumption and costs are not in the consumers focus, on the other hand, it shows that water services are mostly taken for granted. A higher transparency and information level could help to sensitise customers to the good performances and the utilities' challenges. Especially, if ongoing demographical changes make tariff, and therefore price, adaptations necessary, it is important to get the consumer involved. Moreover, clearer requirements on the (economically) appropriate and (legally) correct calculation of prices and fees would help to avoid uncertainties amongst utilities and to mitigate public debates.

For Germany, it can be stated that competition via comparison should be further intensified, because it is a kind of competition, which is not hindered by the regulatory, legal framework. Moreover, the creation of an obligation of higher competition and the development of industry standards should be discussed (see Ottilinger, 2011: 26).

This can help to enlarge the sample, improve the nationwide comparability and achieve even more transparency for stakeholders and customers. Further, the intensification of cross-border benchmarking can help with vision of the bigger picture and ensure the avoidance of project stagnation. The latter is also true for enhanced application of process benchmarks (see Cabrera, 2008: 7).

It should be noted, that the long-term tradition of self-governance seems to be a prerequisite for the proper functioning of the non-regulated water sector in Germany. All in all, the German self-organised water market model works well under the specific German conditions and it can also be expected to work under future challenges. Nonetheless, it will require further efforts, development and above all open-mindedness of all stakeholders for new options and measures in the water market to maintain the current excellent level.

7.10 REFERENCES

Abwasserverordnung (1997). Verordnung über Anforderungen an das Einleiten von Abwasser in Gewässer (Abwasserverordnung – AbwV) from 21st March 1997, last update 2nd May 2013 (BGBl. I S. 973).

Alegre H., Hirner W., Melo J. B. and Parena R. (2000). 1st edn, Performance Indicators for Water Supply Services. IWA Publishing, London. 146p.

BDEW – Bundesverband der Energie- und Wasserwirtschaft (2010). Eckpunkte einer Wasserentgeltkalkulation in der Wasserwirtschaft. https://www.bdew.de/internet. nsf/res/1996251221D35BF5C12578A40044E677/$file/100617_Eckpunktepapier_ Praxisbeispiel_HP.pdf (accessed 3 April 2014)

BDEW – Bundesverband der Energie- und Wasserwirtschaft (2011). Wasserfakten im Überblick.

BDEW – Bundesverband der Energie- und Wasserwirtschaft et al. (2011). Profile of the German Water Sector – 2011. http://www.bdew.de/internet.nsf/id/9F6D4ED10B7 4BCAEC12579570033822C/$file/111129_Profile_german_water_2011_long_.pdf (accessed 23 March 2014)

BDEW – Bundesverband der Energie- und Wasserwirtschaft (2012). Benchmarking: "Learning from the best" – Comparison of performance indicators in the German water sector. http://www.bdew.de/internet.nsf/id/86735E87EB0C1177C12579EB 004E1B9F/$file/120301_BDEW_Wasser_Benchmarking_Broschuere_engl.pdf (accessed 23 March 2014)

BDEW – Bundesverband der Energie- und Wasserwirtschaft; VKU – Verband kommunaler Unternehmen (2012). Leitfaden zur Wasserpreiskalkulation – Gutachten Kalkulation von Trinkwasserpreisen.

BDEW – Bundesverband der Energie- und Wasserwirtschaft (2013). Wasserfakten im Überblick. http://www.bdew.de/internet.nsf/id/C125783000558C9FC125766C0003 CBAF/$file/Wasserfakten%20-%20%C3%96ffentlicher%20Bereich%20August%20 2013.pdf (accessed 23 March 2014)

Berg S. (2010). Water Utility Benchmarking – Measurement, Methodologies and Performance Incentives. IWA Publishing, London.

BGW/DWA – Bundesverband der deutschen Gas- und Wasserwirtschaft/Deutschen Vereinigung für Wasserwirtschaft, Abwasser und Abfall (2008). Wirtschaftsdaten

der Abwasserentsorgung. http://www.bdew.de/internet.nsf/id/DE_Wirtschaftsdaten_
der_Abwasserentsorgung_2005/$file/Wirtschaftsdaten-Abwasser%202005.pdf
(accessed 23 March 2014)

Boscheck R. (2002). European water infrastructures: regulatory flux void of reference? – The
cases of Germany, France, and England and Wales. *Intereconomics*, **37**(3), 138–149.

Bundesministerium für Gesundheit (2011). Bericht über die Qualität von Wasser für den
menschlichen Gebrauch (Trinkwasser) in Deutschland, Berlin.

Cabrera Jr. E. (2008). Benchmarking in the water industry: a mature practice? *Water Utility
Management International*, **3**(2), 5–7.

Cabrera E., Dane P., Haskins S. and Theuretzbacher-Fritz H. (2011). Benchmarking Water
Services – Guiding Water Utilities to Excellence. IWA Publishing, London.

Carvalho P., Marques R. C. and Berg S. (2012). A meta-regression analysis of benchmarking
studies on water utilities market structure. *Utilities Policy*, **21**, 40–49.

Cronauge U. (2003). Kommunale Unternehmen: Eigenbetrieb – Kapitalgesellschaften –
Zweckverbände. Erich Schmidt Verlag, Berlin.

Daiber H. (2010). Die Entscheidung des Bundesgerichtshofes vom 2. Februar 2010,
"Wasserpreise Wetzlar" – neuere Entwicklungen des Wasserkartellrechts. *gwf-Wasser
Abwasser*, **151**(3), 226–235.

Deutscher Wetterdienst (2014a). German Climate Atlas – Air Temperature. http://www.dwd.de/
bvbw/appmanager/bvbw/dwdwwwDesktop?_nfpb=true&_windowLabel=dwdwww_
main_book&T179400190621308654542636gsbDocumentPath=&switchLang=en&_
pageLabel=P28800190621308654463391 (accessed 23 March 2014)

Deutscher Wetterdienst (2014b). German Climate Atlas – Precipitation. http://www.dwd.de/
bvbw/appmanager/bvbw/dwdwwwDesktop?_nfpb=true&_windowLabel=dwdwww_
main_book&T179400190621308654542636gsbDocumentPath=&switchLang=en&_
pageLabel=P28800190621308654463391 (accessed 23 March 2014)

Dierkes M. and Hamann R. (2009). Öffentliches Preisrecht in der Wasserwirtschaft. Nomos
Verlag, Baden-Baden.

DVGW – Deutscher Verein des Gas- und Wasserfaches e. V. (2009). Climate Change and Water
Supply.   http://www.dvgw.de/fileadmin/dvgw/wasser/ressourcen/info_climatechange_
en_100525.pdf (accessed 23 March 2014)

European Commission (1999). European economy – liberalisation of network industries.
Reports and Studies, **4**.

EUWID (2011). Umfassende Wassernutzungsabgabe statt Abwasserabgabe und
Entnahmeentgelt? *EUWID Wasser Special*, **2**.

Ewers H.-J. et al. (2001). Optionen, Chancen und Rahmenbedingungen einer Marktöffnung
für eine nachhaltige Wasserversorgung, BMWi-Forschungsvorhaben 11/00, Endbericht
Juli 2001.

Garcia S., Guérin-Schneider L. and Fauquert G. (2005). Analysis of water price determinants
in France: cost recovery, competition for the market and operator's strategy. *Water
Science and Technology: Water Supply*, **5**(6), 173–181.

GG – Grundgesetz für die Bundesrepublik Deutschland (1949). Grundgesetz from 23rd
May 1949 (BGBl. S.1), last update of 29th July 2009 (BGBl. I S.2248).

Hein A. and Merkel W. (2010). Process benchmarking in drinking waterworks in Germany.
gwf-Wasser Abwasser, **151**(Int. S1), 64–70.

Hirner W. and Merkel W. (2005). Kennzahlen für Benchmarking in der Wasserversorgung.
wvgw-Verlag, Bonn.

Hirschhausen C. von et al. (2010). Wasser: Ökonomie und Management einer Schlüsselressource. Deutsches Institut für Wirtschaftsforschung, Berlin.

Hoffjan A. and Müller N. A. (2011). Contemporary Market Structure and Regulatory Framework, report within the EU Research Project TRUST. http://www.trust-i.net/downloads/index.php?iddesc=34 (accessed 26 March 2014)

Hoffjan A., Müller N. A. and Reksten H. (2014). Enhancing Competition and Efficiency in the Urban Water Industry. Internal Report within the EU Research Project TRUST.

Libbe J. (2014). Zukunftsfähige Wasserwirtschaft zwischen Zentralität und Dezentralität, 47. Essener Tagung für Wasser- und Abfallwirtschaft, pp. 6/1–6/2.

Lotze A. and Reinhardt M. (2009). Die kartellrechtliche Missbrauchskontrolle bei Wasserpreisen. *Neue Juristische Wochenzeitschrift*, **62**(45), 3273–3278.

Mankel B. (2002). Wasserversorgung: Marktöffnungsoptionen umfassend nutzen. *Wirtschaftsdienst*, **82**(1), 40–43.

Merkel W. (2009). Wasser- und Abwasserwirtschaft: Sicherheit und Qualität bestimmen den Preis, nicht umgekehrt, Tagungsbericht von der Handelsblatt-Jahrestagung 2008. *gwf – Wasser\Abwasser*, **150**(1), 72–79.

Oelmann M. and Haneke C. (2008). Herausforderung demographischer Wandel: Tarifmodelle als Instrument der Nachfragestabilisierung in der Wasserversorgung. *Netzwirtschaften & Recht*, **5**(4), 188–194.

Otillinger F. (2011). Benchmarking – was hat es bislang bewirkt und wie geht es weiter?. *energie\wasser praxis*, **62**(5), 24–26.

Petry D. and Castell-Exner C. (2012). Current issues and challenges addressing the German drinking water sector. *Bluefacts – International Journal of Water Management*, **3**, 16–20.

Reif T. (2002a). Preiskalkulation für eine moderne Wasserwirtschaft. *Energie\wasser praxis*, **53**(12), 14–19.

Reif T. (2002b). Preiskalkulation privater Wasserversorgungsunternehmen. Wirtschafts- und Verlagsgesellschaft Gas und Wasser mbH, Bonn.

Statistische Ämter des Bundes und der Länder (2011). Demografischer Wandel in Deutschland.

Statistisches Bundesamt (2011). Statistisches Jahrbuch 2011 – Bundesrepublik Deutschland.

Statistisches Bundesamt (2014). 1000 Liter Trinkwasser kosteten 2013 im Durchschnitt 1,69 Euro. https://www.destatis.de/DE/PresseService/Presse/Pressemitteilungen/2014/03/PD14_110_322.html (accessed 23 March 2014)

The Council of the European Communities (1991). COUNCIL DIRECTIVE 91/271/EEC concerning urban waste water treatment of 21st May 1991.

The Council of the European Union (1998). COUNCIL DIRECTIVE 98/83/EC on the quality of water intended for human consumption of 3rd November 1998.

The European Parliament and the Council (2014). DIRECTIVE 2014/23/EU - Directive on the award of concession contracts from 26th February 2014.

Theuretzbacher-Fritz H., Schielein J., Kiesl H., Kölbl J., Neunteufel R., Perfler R. (2005). Trans-national water supply benchmarking: the cross-border co-operation of the Bavarian EFFWB project and the Austrian OVGW project. *Water Science and Technology: Water Supply*, **5**(6), 273–280.

Trinkwasserverordnung (2001). Trinkwasserverordnung from 21st May 2001 (BGBl. I S.959), last update 14th December 2012, published 2nd August 213 (BGBl. I S. 2977ff.).

Umweltbundesamt (1998). Vergleich der Trinkwasserpreise im europäischen Rahmen. Ecologic – Gesellschaft für Internationale und Europäische Umweltforschung, Berlin.
Umweltbundesamt (2010). Water Resource Management in Germany – Part 1 – Fundamentals.
VKU – Verband kommunaler Unternehmen (2011). Fragen und Antworten – Wasserpreise und Gebühren. http://www.vku.de/service-navigation/presse/publikationen/fragen-und-antworten-wasserpreise-und-gebuehren.html (accessed 23 March 2014)
WHG – Wasserhaushaltsgesetz (2009). Gesetz zur Ordnung des Wasserhaushalts from 31st Juli 2009 (BGBl. I S.2585) last update 7th August 2013 (BGBl. I S. 3154).

Chapter 8

Assessment of water services from the perspective of multilateral organisations.
The aquarating experience

Matthias Krause and María del Rosario Navia
Banco Interamericano de Desarrollo

8.1 INTRODUCTION

Governments, civil society organisations and development partners the world over are striving to improve water and sanitation services for families and businesses in order to guarantee universal access to safe, sustainable, efficient services. Reliable, quality information about the situation of the service and its performance is fundamental for the people responsible for designing policies and business managers providing services to be able to improve and design effective policies and projects to overcome any shortcomings. Development organisations, who partner up with Governments, service providers and civil society organisations also depend on sound information to design, finance and successfully execute their projects. With best practice and performance rules, reliable information can become a powerful tool to encourage and direct efforts by the different interested parties, to mobilise human and financial resources, and to achieve improved services. Enforceable rules under state regulations and voluntary rules are two complementary approaches that can be used when creating a virtuous circle of these characteristics. The aim of this article is to discuss this approach in greater detail and to present and analyse a recent example: AquaRating.

AquaRating is a new performance assessment system for companies who provide water and sanitation services on behalf of a certified independent party. The authors believe that this new instrument could make a substantial contribution when establishing performance rules for universal use in providing services in urban areas, assessing practices and performance of service providers on the basis of reliable information, and to afford guidance for improvement proposals to providers and people responsible for defining policies. The main idea behind the AquaRating approach is very similar to the underlying ideas of modern regulatory approaches: to provide users with a better service by establishing rules to govern service provision, encourage measuring the service and supervision in accordance with said rules, and also to encourage service providers to involve themselves in a learning and improvement process. However, there are also considerable differences between the two approaches. The most important difference is that whereas AquaRating is a voluntary system which in principle is applicable all over the world, regulating water and sanitation services – as it is conceived in many parts of the world – is a mandatory system ultimately approved by a sovereign state (or a union of sovereign states) and is only applicable within the relevant frontiers. Therefore, a voluntary system such as AquaRating is clearly complementary to state regulation.

The authors of this article are members of the team responsible for developing and testing AquaRating and for putting the system into practice. It is a joint initiative between the Inter-American Development Bank (IDB) and the International Water Association (IWA). According to the objectives of these two organisations, the main reason for setting up this initiative was to develop a new system that contributed to improving water and sanitation services for homes and businesses in terms of access, quality, efficiency, transparency and sustainability, and which is useful for service providers, policy-makers, finance institutions and development partners.

The rest of this document is structured as follows: The first section deals with motivation to encourage the use of reliable information when assessing the performance of service providers within the context of challenges concerning regulation. The second section describes the approach and structure of AquaRating, along with experiences in field tests. The third section outlines the conclusions.

8.2 CHALLENGES CONCERNING REGULATION AND THE VALUE OF ASSESSING PERFORMANCE AND RELIABLE INFORMATION

The aims of regulating water and sanitation services are basically to protect consumers from abusive monopolistic power by the service providers – whether through excessive prices or deficient services, to promote public health, to enforce compliance with public services when they exist, and to protect the water resources and the environment. The specific characteristics of water and sanitation service provision that justify regulation of this industry – the natural monopoly, the external effects and their relationship to human rights – are well documented in literature (see e.g. Krause, 2009, Ch. 2).

Regulation all over the world faces a series of challenges when attempting to achieve its objectives (see Rouse, 2007 for a thorough review). These challenges range from designing a suitable global regulatory approach, owing to the specific structure and governance of the water industry in the specific country, to guarantee independence and credibility of the regulatory decisions, promoting participation by users and hiring personnel with sufficient knowledge and experience, and then on to standardising regulations in terms of tariffs, environment and service quality whilst applying regulatory techniques that are effective for service providers whether they are state or private companies (see Berg, 2013 to consult a recent contribution to the latter). Governments and companies have found different approaches to put regulation into practice, although it is true that in many countries, particularly in rural areas, regulation of water services is still non-existent, at least in practical terms (see Marques, 2010 for a world comparison on the state of regulation). In general terms, there are two kinds of regulatory approaches. Firstly, self-regulation of the sector, voluntarily, and secondly enforceable state regulation. With the second category we are able to differentiate between contracted regulation and regulation through a specialised regulatory organisation.

In this document we differentiate between "voluntary self-regulation" and "absence of regulation". In the first case the service providers voluntarily undertake to comply with certain rules or performance objectives and implement some kind of system to submit reports and a supervision system permitting public scrutiny. In the second case, the performance rules or objectives (whether mandatory or voluntary) are non-existent or only exist on paper, but in practice are not implemented and no reports are produced or supervision takes place.

A key piece of any regulatory system is reliable information about the levels of service and management practices implemented by the service providers (Schäfer *et al.* 2012). Nevertheless, in practice this is the Achilles tendon of many regulatory systems because the regulator does not find it easy to obtain reliable information. For the same reason which justifies economic regulation of water supply services – the monopolistic nature of the service – no information is available about performance and costs, and many service providers do not have any clear incentives to share this information. This means that in most countries, systems must firstly be established for submitting reports and supervising service levels and management practices, and the reliability of the information must be verified and improved before regulation can become truly effective.

Accurate, credible, comparable information about performance by service provider companies is not only important for regulators, but it is also crucial for directors and managers of these companies, and also for investors, finance entities and development partners. For example, if said information is available, the company's directors will be able to use it to identify areas for improvement, for example in a benchmarking environment, to give priority to improvement action and justify the need for investment. The definition of benchmarking or comparative assessment (Cabrera *et al.* 2011) is as follows: "Benchmarking is a tool for performance improvement through a systematic search and adaptation of best practices". For a financial institution for development, such as IDB, it means that identifying and preparing projects is much simpler and less costly because it is necessary to compile and assess less technical, financial and institutional data ad hoc about the public services and also because there is reliable information about the performance of the service providers which can be used as a reference for the project-specific information that would need to be compiled regardless.

In view of the numerous advantages of common standards for performance assessment of service providers, IDB and IWA joined forces to develop the AquaRating system. This new approach is based on three principles: (i) to establish a reference system for universal use to measure performance reflecting best international practices in terms of water and sanitation providing services in urban areas; (ii) to offer service providers a system of voluntary assessment of their performance by a certified, independent party ("AquaRating") in accordance with this reference system based on documented, verified information which is therefore reliable; and (iii) to give priority to service providers, their needs and motivations by permitting them to obtain specific advantages from AquaRating and thus making the tool more attractive to use. The combination of these three elements is also what makes AquaRating different from other currently existing initiatives with similar goals, such as the benchmarking projects managed by regulators, service providers or multilateral organisations. The AquaRating approach and experience are described in the following section.

8.3 AQUARATING APPROACH AND EXPERIENCE

This section is mainly an adaptation of the introductory chapter of the Inter-American Development Bank (2013, pages 3–9).

8.3.1 Approach

The currently available pilot version of AquaRating comprises 113 assessment items divided into 8 areas, each of which is assigned a score, which in turn are included in a single aggregate mark ranging from 0 to 100 rating the service provider. Approximately half of these items are quantitative indicators of more traditional performance, whereas the rest of the items consist of new measures of best practices specifically developed for this project. Full compliance with all the best practices and achieving the most demanding levels for the indicators would entail excellent service provision and therefore a maximum score of 100 points.

Including best practices stems from several aspects of the three design principles mentioned at the end of the previous section. On the one hand, best practices are less context sensitive. By assessing the way the services are managed, and not only the numerical results from said management, the geographical, economic and social conditions where providers operate do not have such a strong influence as they would in an assessment exclusively based on numerical results. On the other hand, practices permit establishing a future forecast for the service. For example, a service provider who today reports reasonable results, but has shortcomings in good practices in a critical area, will probably suffer sustainability problems. Finally, another virtue of the best practices used in AquaRating, and which was emphasised by the service providers who took part in developing and testing the system, is that they are a guide in themselves, since the practices that are not currently implemented can become short- and mid-term objectives for the service provider.

AquaRating management will be entrusted to an independent entity appointed by IWA, according to a financial self-sufficiency model in which the service providers or a sponsor defray the auditing costs. This approach appears to be desirable and necessary for the system to be sustainable over time. Furthermore, it could even be said that this feature is a kind of fire test for the system: if a provider or a sponsor is prepared to defray the costs of the rating service, it means that the service has true value for the provider. The information compiled from the audit will be confidential and the provider will decide whether or not to make it public. In order to guarantee credibility and to ascertain the reliability of the information supplied by providers, all the information employed to calculate the final score must be supported by documents and validated through an audit. This audit, which will take place after a previous self-assessment stage by the provider, will be carried out by independent auditors accredited by AquaRating.

8.3.2 From the perspective of the service provider, the following advantages are obtained from participating in a system such as AquaRating

Reputation and acknowledgement by interested parties: The service provider can report the rating to certain interested parties or make it public as part of a communication strategy.

Service providers can identify improvement opportunities and establish action plans for the identified areas.

Access to financial and training resources. With the assessment afforded by AquaRating, a service provider will have sufficient information to improve access to Government or Development Bank financial and training resources, or even private banks.

A service provider who knows the current performance level and develops a solid improvement plan for water and sanitation providing services with enhanced quality, efficiency, sustainability and transparency.

8.3.3 Structure

In order to offer truly thorough assessment, AquaRating covers all aspects of management in service providing companies through the following eight areas (Figure 8.1):

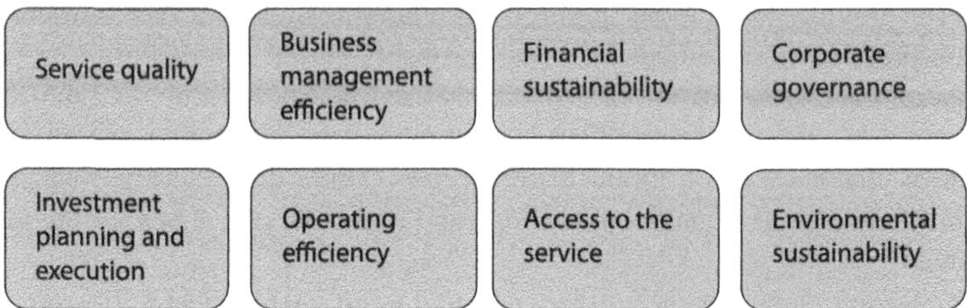

Figure 8.1 The eight areas of management in service providing companies that are covered in AquaRating.

Each of the assessment areas are divided into sub-areas, which in turn are divided into assessment items. For each assessment item, a score from 0 to 100 is calculated according to the information supplied by the service provider through an information system. The scores are then aggregated in sub-areas and finally at global level. The service providers must furnish documentation supporting the information they enter.

In order to obtain a score for each item, the system standardises the results for each one. The reliability of the information source for each of these items is also assessed, and the score is modified according to said reliability by correction weighting. This process is illustrated with the example of the *Drinking Water Quality (CS1)* sub-area, which is one of the four sub-areas into which the score for *Service Quality* is divided. The assessment items contained in said sub-area are shown in Figure 8.2:

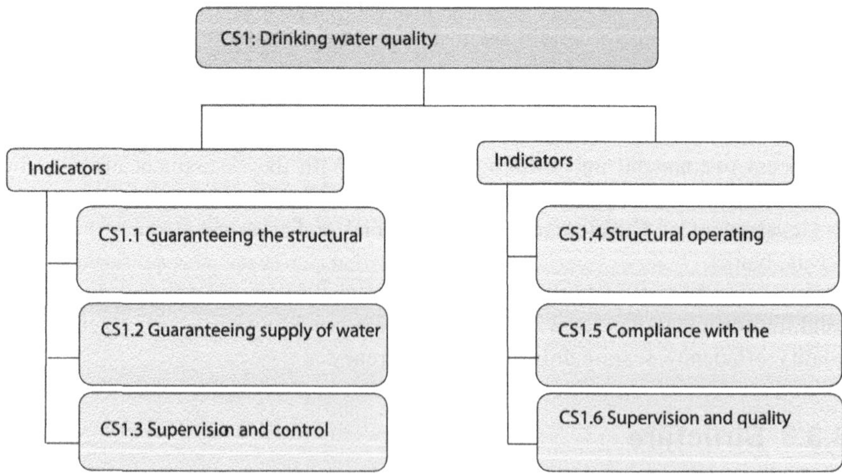

Figure 8.2 Example of the *Drinking Water Quality (CS1)* sub-area within the AquaRating structure.

In the case of the list of practices, the score is standardised by assigning relative importance to each of the practices on the list (weight). For example, for assessment item CS1.1 *Guaranteeing structural capacity for treatment and supply*, the list of best practices and the corresponding weighting are shown in Table 8.1. If all the practices are met, a score of 100 is assigned to each assessment item, whereas partial compliance will entail a proportional deduction from the score depending on the specific weight of the item that is not complied with.

Table 8.1 Example of how a standardised score is achieved through assigning relative importance to practices (for assessment item CS1.1 *Guaranteeing structural capacity for treatment and supply*).

Practices	Weights
1 There are protection measures when capturing raw water (signs, protection perimeters, fencing, etc.) for the sources of raw water assessed in the system.	1
2 There are action protocols for "preventive maintenance" at treatment stations and records thereof.	1
3 There are protocols for "corrective maintenance" action at treatment stations and records thereof.	1
4 There are automated processes for operating in the absence of personnel at the treatment stations (or available personnel 24 hours per day).	3
5 There are protocols for analysis and resolution of breaches of "applicable regulations" concerning water quality with notification reported to the competent authority.	2
6 There are safety plans for "contingencies" concerning water quality.	1
7 There are protocols to guarantee water quality when commissioning new supply sources (wells or surface water sources).	1
8 There are protocols to guarantee water quality when commissioning new infrastructures.	1

In the case of the indicators, their values are calculated using a formula consisting of variables. Standardising the score is carried out by means of an objective function marking the desirable values for each indicator. For example, when assessing CS1.5 *Compliance with drinking water regulations*, the formula is:

([CS1-V3] / [CS1-V2]) * 100

Where the variables are:

[CS1-V3] = Number of population for which the rules have been met that are applicable to the water samples.

[CS1-V2] = Number of population who receive the service in the geographical area to be scored.

In Figure 8.3 the function of standardising is shown with the value of the indicator on the horizontal axis and the standardised AquaRating value on the vertical axis. The compliance percentage with the drinking water regulations (% complying samples) gives a standardised value of 0 for 80% of correct samples. A total of 92% of the samples report a standardised value under 30, whereas 99% of samples permit reaching the maximum score of 100.

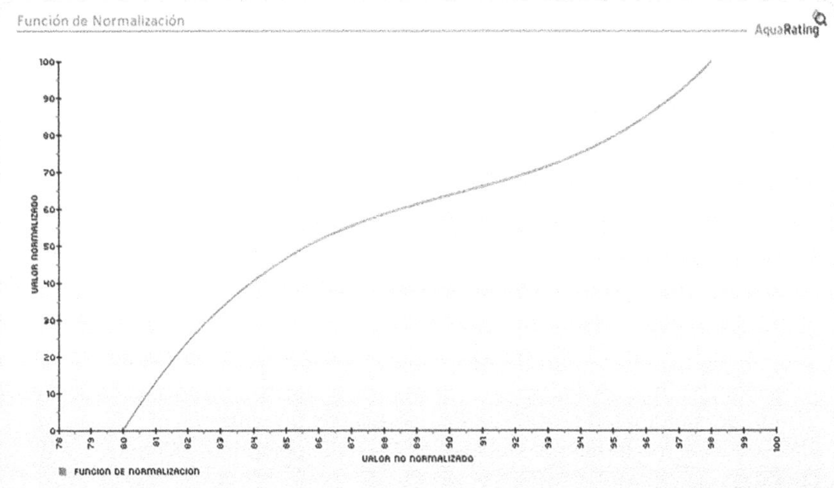

Figure 8.3 The function of standardising. Shown with the value of the indicator on the horizontal axis and the standardised AquaRating value on the vertical axis.

The system assigns a reliability correction factor for practices and indicators scores, depending on the available information. For example, the correction factors for variable [CS1-V3] are shown in Table 8.2.

Table 8.2 Example of the correction factors for variable [CS1-V3].

Reliability	Correction Factor
No records.	0
Controls and sample analyses were carried out in unsigned records without quality control.	0.33
Controls and sample analyses were carried out in signed records, with traceability and quality control.	0.8
Controls and sample analyses were carried out in signed records, with traceability and quality control. And a reliable system was established to link samples with the population or with the relevant properties.	1

The score for each of the areas is obtained by adding the weighted score of each of the items hierarchically beneath them (Figure 8.4). Hence, at each level of the AquaRating system, a score is assigned ranging from 0 to 100. These scores are weighted at the same level with pre-established weights, so that each area and sub-area is assigned a score from 0 to 100, resulting from combining said items.

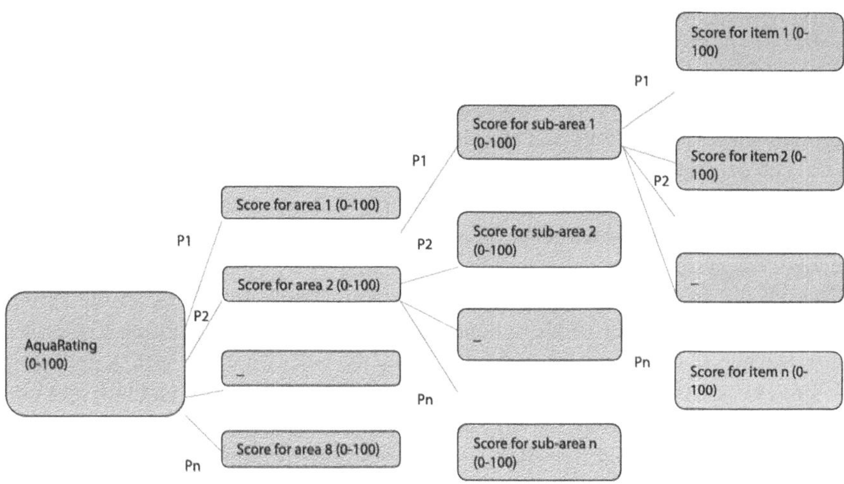

Figure 8.4 AquaRating score hierarchy.

Self-assessment by service provider affords an initial rating, which is subsequently verified and validated by an external auditor accredited by AquaRating. Depending on the audit report, the AquaRating entity prepares a qualification report for the service provider.

8.3.4 Field test development and experience

Development of this tool followed a dynamic process with participation by and ideas from service providers, international experts and other multilateral organisations through different seminars, group meetings by subjects and direct conversations.

When the AquaRating pilot version was completed, it was tested at 13 service providers in Latin America and Spain with the aim of improving the AquaRating tool and adapting it for use in real situations. More specifically at Agua y Saneamientos Argentinos S.A. (AySA), Argentina; Companhia de Saneamento Básico do Estado de São Paulo (SABESP), Brazil; Empresas Públicas de Medellín (EPM), Colombia; Empresa de Acueducto y Alcantarillado de Pereira S.A., Colombia; Aguas Andinas, Chile; Aguas de Cartagena S.A., Colombia; Servicios de Agua y Drenaje de Monterrey I.P.D (SADM), Mexico; Empresa Pública Metropolitana de

Agua Potable y Saneamiento de Quito (EPMAPS), Ecuador; Obras Sanitarias del Estado (OSE), Uruguay; Corporación del Acueducto y Alcantarillado de Santiago (CORAASAN), Dominican Republic; Aqualia, Gestion Integral del Agua S.A., Servicio Municipal de Aguas de Almería, Spain; Aguas Municipalizadas de Alicante, Spain; and Empresa Municipal de Agua de Córdoba, S.A. (EMACSA), Spain. The participating providers were carefully selected to cover a wide range of regulatory frameworks, sizes, ownership methods, management models and technical specifications. Some of the service providers were audited in order to also test the auditing process.

The impressions of service providers and auditors were systematically compiled through surveys, meetings and the incident system integrated in the AquaRating tool. The principal conclusions from the tests confirmed that: i) the on-line tool is easy to manage; ii) AquaRating can be used in different contexts; iii) the providers value the knowledge featured in the system; iv) the providers value a universally recognised system that can reliably certify their performance; v) the providers discovered aspects about their own organisations thanks to the use of the system and identified areas for improvement; vi) some providers transformed what had been learned in improvement strategies, using the rules established in AquaRating as references; vii) the information furnished by providers can be audited; and viii) some consider AquaRating a useful tool to obtain access to financing.

The tests also served to collect ideas about how to improve the strategy. The main improvements are: i) to speed up procedures to document and load supporting information; ii) to speed up the auditing procedures; iii) to improve the definition of the supporting documents; iv) to adjust the definitions of the assessment items and update the software and documents related to the tool to record all changes; and v) to improve the browsing functions of the software.

8.4 CONCLUSIONS

The ultimate goals of state regulation in terms of water and sanitation services are to guarantee that all homes and business in certain areas of responsibility, actually receive good quality services in an efficient manner, to promote public health and protect water resources. In order to achieve these goals, regulations must establish significant rules for service levels and management practices, and also obtain reliable information in order to supervise service providers and attain compliance with the regulations. A rating system such as AquaRating is a complementary tool for a service provider in a regulated environment, since it promotes improvements in water and sanitation services in terms of access, quality, efficiency, transparency and sustainability. In an environment marked by fairly underdeveloped state regulation, a system of this kind could serve as voluntary "self-regulation".

AquaRating features innovative, distinctive items that make it unique, thorough and reliable. These items are: i) universality: it can be used by urban service providers anywhere in the world and it is applicable in different contexts;

ii) thorough assessment: it assesses all the important aspects of management by service providers and not only includes performance indicators, but also practice indicators, so that assessment covers a wider scope; iii) reliability: AquaRating requires service providers to furnish documents supporting the information they have supplied. This supporting information and documents are verified by an external auditor and AquaRating is certified by an independent third party (AquaRating Entity). This is very important since it has been recognised that the quality of the data is crucial for credibility of any performance assessment system; and iv) information to improve the provided services: it is capable of assessing current performance and identifying the potential for improvement.

Since AquaRating is a universal system that provides a standard assessment platform, service providers are not the only parties to benefit from it, as Governments, civil society organisations, development partners and finance institutions can also find it a perfect tool to help direct decision-making when giving priority to financing and efforts.

8.5 REFERENCES

Banco Interamericano de Desarrollo (2013). Sistema de Calificación de Prestadores de Agua y Saneamiento AquaRating. Documento Técnico, Versión 1.1.2. 31 de julio de 2013. Documento interno elaborado por: E. Cabrera jr., F. Cubillo C. Díaz J. Ducci and M. Krause.

Berg S. V. (2013). Best Practices in Regulating State-Owned and Municipal Water Utilities. Comisión Económica para América Latina y el Caribe de las Naciones Unidas (CEPAL), Santiago.

Cabrera E., Jr., Dane P., Haskins S. and Theuretzbacher-Fritz H. (2011). Benchmarking Water Services: Guiding Water Utilities to Excellence. IWA Publishing, London.

Krause M. (2009). The Political Economy of Water and Sanitation. Routledge, New York, London.

Marques R. C. (2010). Regulation of Water and Wastewater Services: An International Comparison. IWA Publishing, London.

Rouse M. J. (2007). Institutional Governance and Regulation of Water Service: The Essential Elements. IWA Publishing, London.

Schäfer D., Goertler A., Gerhager B. and Gerlach E. (2012). Through the Looking Glass: Designing and using monitoring systems for effective regulation and performance management in urban water and sanitation service provision. Regulation Brief No. 2. Eschborn: Giz.

Chapter 9

Can a regulator contribute to resolving the main problems of the urban water cycle in Spain?

Enrique Cabrera
ITA, Universitat Politècnica de València

9.1 INTRODUCTION

Urban water services have evolved enormously since they were first born (over a century ago in big cities) and are a key service in the quality of life of citizens today. But some important dysfunctions have appeared in recent decades because the changes brought on with time (such as maximising the guarantee of water supply) are performed at a much greater speed than the implementation demanded by the new legal ordinance. In other words, as the complexity of management and demands of citizens has grown, the administration has not adapted the regulatory framework (in the case of Spain it has not even been created) to this new context. This article proves this statement. And to do so, we shall take a brief look at history. We are reminded of the irrefutable objective to achieve in performing these services, namely to provide the service sustainably, with maximum quality and at the lowest possible cost. Finally, the complex challenges these services face are listed, followed by an analysis of the current dysfunctions and the changes that will be required to correct them as time goes by. This paper, which includes a reflection about two continuously ongoing topics (the universal right to water and the public/private management debate), concludes with an obvious question within the context of this book: To what extent can a regulator contribute to resolving these problems? It is worth mentioning that it was precisely these very problems that in the past brought about profound reforms in other countries, as seen in the first chapters of this book.

Urban water supply in Spain, as we know it today, was born in the second half of the 19th century. The quality of life of Spanish citizens took an enormous leap forward through these systems, in fact, in terms of health it marked a before and after. It is hardly surprising then, that implementation spread very quickly and around the middle of the 20th century nearly all urban centres had this service. And as many of the pipelines that were originally installed are still in service today, the first reflection appears to be unavoidable. If the average age of pipes is around 50 years, many kilometres of this network have exceeded their useful life. This is by no means a trivial matter. Investments at the time were made out of need, and with the dream of having a vital service – having running, drinking water in the taps of homes. But today, motivation to renew them is much lower. As it was the administration who developed the major water infrastructures, the belief is that the administration should also upgrade them. And users should only defray a part of the costs. It has therefore changed from a need and a dream to a demand, with citizens having practically no responsibility, which is what conditions their attitude today.

Because we are used to having this commodity, citizens have converted it into a right. Nevertheless, people are not aware, or have not stopped to think about the fact (because nobody has bothered to explain it), that pipes need to be renewed, and usually complain when they are replaced, above all because it is inconvenient. Complaints become even louder when this entails a tariff increase. Basically what was once a dream, and therefore was paid for without qualms, willingly bearing the inconveniences that were necessary, is seen today as an acquired right and people today are not willing to accept either extra charges or inconveniences, which leads us to a second question: Is it necessary to teach citizens to understand that only with new investments and occasional inconveniences (although trench-free technology today means networks can be renewed and inconveniences minimised) can the quality of this service be guaranteed and can it survive over time?

The birth of these services, as there could be no other way, was driven by the towns themselves. Defended by the economy of scale, with a very reasonable investment, water could be taken to the taps in people's homes. In some urban areas, particularly in residential complexes, time-wasting and hesitation by the administration meant that private initiatives took responsibility for developing the service. Over time, and in view of the demands to guarantee the quality of water in taps, town halls have bought most of the privately owned systems, and therefore today, with only a few exceptions, these strategic urban infrastructures are municipal property.

And of course, what was in fact being consolidated in the early decades of the 20th century (the responsibility of town halls to supply citizens with drinking water) had to be regulated. This task was undertaken in 1985. It was then that the Local Regime Regulatory Rules law was enacted (Law 7/1985, of 2nd April), which establishes (in articles 25.2.1 and 86.3) that one of the mandatory municipal services is the supply of water. And of course this is only logical since urban water networks are by definition local, although several towns can join together for general distribution of the resource, which is known today as supply grids (*suministro en alta*). This logic, although with some major exceptions as seen in previous chapters, prevails in nearly all of the world. There is nothing to object against concerning this matter. Municipal responsibility, although not universal, is widely generalised, as is only logical.

But the growing technical and economic complexity inherent to these services (guaranteed drinkability of water in taps, minimising the needs for the required natural resources – water and energy, returning used water to the natural environment at the quality required under the demanding European directives imposed today, and basically complying with the quality standards of the service itself) very often overwhelms the technical capacity of townships, particularly the smaller ones. These deficiencies have been largely resolved through support by higher institutions, such as the provincial councils in the case of smaller towns and villages. But after decades of municipal management of these systems, it appears reasonable to see how things have been going in order to reinforce the positive side of things and make changes where there is room for improvement. This is a particularly relevant reflection within a new framework; a new framework in a new era (with the start of the economic crisis at the end of the previous decade) forcing all involved parties to be much more efficient than they have been until now.

9.2 THE URBAN WATER CYCLE: OBJECTIVES

To reach an objective with the minimum possible effort it is necessary to know the starting point and the finishing point. in order to define the former, a three-stage, technical diagnosis needs to be conducted:

- *A water audit to to accurately determine the efficiency of the system.* A metrics audit, using representative indicators. (This is not so much about performance percentages, but rather litres/supply/day, or cubic metres/kilometres/hour). This audit should permit separation of real losses (leaks) and apparent losses (measurement errors and theft of water).
- *An energy audit in order to calculate the energy intensity (kWh/m³) in the different stages of the urban cycle, from water capture to distribution.* The calculation should include the energy lost through leaks (which is a loss of two resources: energy and water).
- *An economic audit.* The urban water cycle as a whole has a heritage value that needs to be preserved over time. It is determined on the basis of the initial cost of the infrastructures comprising the system, and afterwards, on the basis of the average life of the different infrastructures, determining the current residual value. A maintenance and investment programme then needs to be established with the objective of at least guaranteeing preservation, and if possible improvement, until the conditions of the infrastructure reach an acceptable level. All these costs need to be passed on to the water tariffs, with the structure of these tariffs being adapted to the social circumstances of the town.

The methodology to perform this diagnosis with the three audits is universal, but the results obtained for each urban area will undoubtedly be very different. Nevertheless, all of them share the aforementioned ultimate goal: to provide a quality service that is sustainable over time at the lowest possible cost. There are three ingredients then: quality, reasonable cost and sustainability. The first is defined through the quality standards of the service to be provided. The standards in Spain, exclusively referring to the quality of the water in taps, are well-defined. The others may exist locally (defined in service outsourcing contracts) but there are no general directives. The second, namely the minimum cost, depends to a large extent on the circumstances of each particular service. It is not the same having a source of good quality water from a natural spring, as having to desalinate seawater. Topography also plays its part. Another key factor is an economy of scale. There are many indicators to quantify this: number of delivery points per kilometre of network, kilometres of network per one thousand inhabitants, etc. Therefore, in this second diagnosis it is necessary to resort to benchmarking, a widely discussed subject in this book. But there are mechanisms to assess the economic management of the system, where the training of the decision-makers plays a decisive role.

The third ingredient, sustainability, despite being the one that is most repeated, is not always the best understood. Therefore, and before moving on to analyse this term, it is opportune to clarify just what is behind this cliché coined by the United Nations Brundtland Commission in 1987. The definition given in the final report "Our Common Future" (Brundtland, 1987) is "*Sustainable development meets the needs of the present without compromising the ability of future generations to meet their own needs*". In the case we are dealing with, and *mutatis mutandi*, it must be said that providing this service sustainably means doing it in such a way that does not compromise it for future generations, at least with the same quality. And in order to do so, we will have to use clean supply sources, infrastructures that permit it to be in good condition and all the foregoing performed in line with social awareness. We should not overlook the fact that we are dealing with a basic service, one that is vital for human life. All said, and even though it does not comprise the ultimate goal of this paper, we should not forget that since urban water services are encompassed in a more general context, the concept of sustainability permits very different interpretations, which are still under discussion (Werkheiser & Piso, 2015).

Regardless of the requirements, as mentioned previously, still under discussion, the concept of sustainability should be framed in a three-dimensional area by three other axes: environmental, economic and social. One way or another the axes have conflicting interests. The environment axis defends the interests of the natural environment (which entails good treatment of water) requiring investments and operating costs, thus affecting the economic axis. A point of balance needs to be found between the three of them. Because in some way, none of them is more relevant than the other two. All of them play a fundamental role, and at the same time are complementary to each other. And the best way to prove this is to remember (in alphabetical order) the significance of each item. Which is as follows:

- *Environmental axis.* The urban water cycle is artificial. Man builds a series of infrastructures to have quality, running water in his taps at home. This means water must be diverted away from its natural cycle, forcing it along an alternative route which must be done with minimum environmental impact along the way. This means doing so taking the least possible amount of water from the natural bodies of water (which requires efficient use) and once it has been used (and therefore contaminated and degraded), returning it to the natural environment at least with the same quality as when it was diverted from its natural route.
- *Economic axis.* The infrastructures comprising the urban water cycle are very expensive and require colossal investments. We are not really aware of this because they have been gradually installed over many decades. But as a whole they comprise heritage of exceedingly high value and will only meet the environmental quality objectives and efficiency objectives if they are in good condition. This is only possible by permanently maintaining and renewing them whenever their useful life runs out, which also entails considerable investments. In practice (Copeland, 2013) the usual renewal periods exceed the average recommended life.
- *Social axis.* Nobody can deny man's right to water and sanitation. This was recognised by the United Nations in July 2010. Consequently, man must be able to have access to a basic daily volume of water at an affordable price for all classes, conditions and circumstances. The social axis however, should not only consider human rights. It should also ensure that tap water is not only drinkable, but should be of a quality allowing citizens not to be forced to drink bottled water (thousands of times more expensive, and much less environmentally friendly). These social objectives should be made compatible with the environmental and the economic objectives. And this is only possible if there is enough money to make the necessary investments, and it is spent wisely.

In short, and emphasis is brought to this point, the social axis involves interests that are inherent to the other two. This is evident through the fact that citizens are more reluctant to assume costs linked to water. But when it is analysed pragmatically, the solution is relatively simple. Effectively (INE, 2015a), in 2013 in Spain, the average annual expense for a citizen for the complete water cycle was € 86.8 (average consumption of 47.45 m³/year at an average cost of € 1.83/m³) which is an affordable amount for most citizens, even in times of crisis. In fact, in the same year the average income per capita was € 22,291 (El País, 2013), and therefore the cost for water and sanitation was less than 0.4%. Any other not so basic service (electricity, gas, telephone) is not essential, and considerably more expensive. And nobody goes without them.

On the other hand, experience shows that if water is not assigned the right price in accordance with the principle of cost recovery, it is not used efficiently. This is an obvious reality which brings to light the absolute necessity of recovering all costs and as far as possible avoiding any subsidies. This statement should not be shocking. This is stated, and not by chance, in the Water Framework Directive (EU, 2000). Basically the average price of water should be calculated by balancing all the costs (including, obviously, besides maintenance, the amortisation of installations in Spain which is generally absent) with the revenue from users. And the social axis should be addressed with gradual tariffs that guarantee the basic water requirements (very low prices, or at least without cost for the most needy classes) and afterwards, establishing gradual steps. In other words, the social and environmental axes need to be moved away from the basic principle of a free market, which usually gives priority to a lower unitary cost with higher consumption (as more is used, less is actually paid). In synthesis, as much access to water must be provided as possible, while taking the least amount of water away from the environment, and to do this efficiency is required.

Figure 9.1 depicts the foregoing, and at the same time conceptualises the overlaps of the axes. The most complex combination of the three; economic - social, is summarised under the term "Fair", with this fairness being achieved through adequate tariffs. The other two terms in the overlaps are easily understandable. For the environmental - economic overlap "Viable" permits the former and for the environmental - social overlap "Habitable" is used, because degradation of the natural environment makes the territory uninhabitable. Living within a dignified framework is one of humanity's social rights. The overlap where the three meet, "Lasting", synthesises the meaning of the term "sustainable". And as is only logical, only governance (Organisation for Economic Co-operation and Development [OECD], 2015) can make the three compatible with each other, although this is a truly complex task.

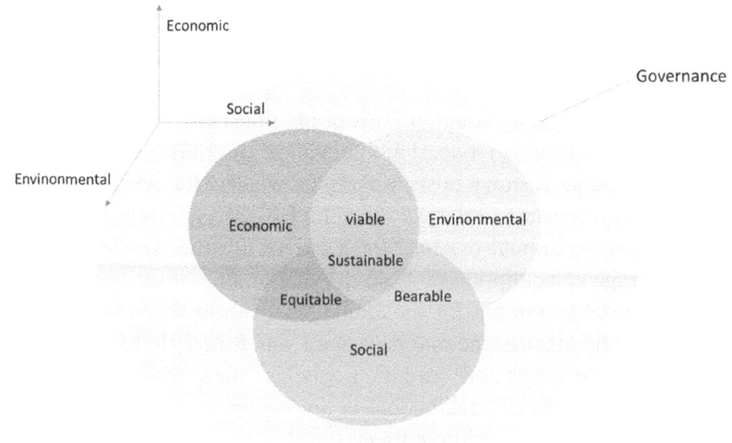

Figure 9.1 Concept of sustainability and application to the urban water cycle.

Having arrived at this point, we are now able to establish the road map to take us to the main final goal – managing quality water at the lowest possible cost. In other words, based on the diagnosis of the system, a realistic road should be identified, but one that is as short as possible, that leads to the final destination – that of social, economic and environmental sustainability of the system.

9.3 THE STATE OF AFFAIRS IN SPAIN AND IN THE WORLD

The current state of affairs in urban water cycles in Spain cannot be generalised, as there are always exceptions to confirm the rule. And less so in a world where, obviously, the scope of possibilities is so very wide. But there are several obvious facts and trends that lead to a number of considerable challenges which managers of these systems need to address. These challenges are listed and discussed in the following lines (the order is not important, since the degree of relevance of each particular challenge depends on each specific case).

9.3.1 Ageing infrastructures

The reasons for this situation of ageing infrastructures have already been discussed. It is relevant to point out that Spain is no exception in this situation. The problem is universal. Hence, in the country that has been leading the world over the last century, namely the United States, this is a subject of growing concern and worry. This can be seen through the number of publications highlighting this problem. In fact, there are numerous reports that deal with the needs of ageing infrastructures. The Environmental Protection Agency (EPA) quantifies these needs every four years. The fifth and latest of these assessments is for 2011, which was published with a delay of two years (EPA, 2013). It is estimated that in the next 20 years, upgrading only the drinking water supply in the USA will require an investment of 384,200 million dollars. Bearing in mind that in 2009 the population was 315 million, the annual investment requirements are estimated at 61 million dollars per million inhabitants per year (around 55 million euros per one million inhabitants).

Other studies in the same country (AWWA, 2012) only focusing on drinking water pipelines, estimate that to upgrade these networks (covering an approximate length of 1,600,000 km) will require an investment of over 2,400,000 million euros in the next 25 years (1.5 million euros per kilometre of pipeline in 25 years). Supposing the requirements are similar (which depends on the condition of the pipelines and growth of networks) these figures could be translated to Spain (46 million inhabitants and 250,000 km of pipeline). The results give average requirements of 33 million euros per one million inhabitants, a value of the same magnitude, although rather more conservative than the result in the EPA study (55 million euros per one million inhabitants). The difference is easily justifiable since the first assessment considers water capture, treatment and storage works, whereas the second only considers the pipelines. Whichever the case, there can be no denying that we are talking about very significant numbers.

No data has been published for Spain (perhaps a study of this kind would be advisable), but, whichever the case, the pending investment will be significant. Bearing in mind that this study is already two years old, and that the infrastructures in Spain are probably in poorer condition than in the USA, the cited amount could be rounded up to 50 million euros per one million inhabitants per year.

In times of economic crisis, with almost unbearable levels of unemployment, this figure should not be shocking. In view of the notable reports, including the one by the Rockefeller Foundation (GFA, 2011), this should be seen as a major opportunity. The report concludes that with an investment of around 45 million euros per year (the figures in the report refer to the annual investment required to maintain the urban water infrastructures per one million inhabitants), complementary activity of 63 million will be generated (a multiplying effect of 1.4) while creating 53,00 jobs – "green jobs" since they will contribute to conserving the natural environment. Hence, we could say that two objectives are achieved with the same action. Another analysis of the same type, by the Associated General Contractors of America (AGC) and the Clean Water Council (NWAC, 2012) reaches more optimistic results. With an investment of 1000 million dollars, 28,500 jobs would be created.

9.3.2 Increasing pollution

Current standards of living have undoubtedly caused severe degradation of the natural environment in general, and specifically of bodies of water. All human activities that require water degrade its quality. If, as has all too often been the case in recent decades, water is returned to the environment untreated, pollution increases. This can be focal (spillage from an industry of city) or diffused, the second caused by arable and livestock farming. Fertilising fields and subsequent leaching of these chemical agents through irrigation water by filtration, reaches aquifers thus causing contamination from arable farming. Pollution from livestock farming is when rain carries slurry into aquifers, with the same results of contamination. Since there are so many sources of pollution, control is much more difficult.

The damage caused by pollution is always higher than the costs involved in correcting it at source. To demonstrate this, the economic balance needs to be conducted well, with a comprehensive approach to the complete water cycle. But as competences are widely dispersed, it is hard to make decisions that maximise general interests. Whichever the case, global economic analyses prove that it is better to treat, and better still to avoid contamination at source. What can only be defined as common sense has been quantified in Denmark and the United Kingdom (Viavattene *et al.* 2011). Different analyses have assessed lost profits by farmers through not fertilising their fields and consequently not polluting aquifers, which were later used for urban water supplies. The cost of treating water contaminated by nitrates was always higher than the loss of income by farmers.

The economic and environmental impact of pollution should always be dealt with globally. And, without a shadow of a doubt, decontamination at source is much cheaper than any other solution, when all the costs are taken into account. The challenge of making progress towards comprehensive management is one of the biggest water policies have to address. Contamination is the main reason, although it is not the only one, to address problems globally, being the only one capable of responding to the current challenges in a globalised world. This is particularly true taking into account the increasingly strict standards Brussels demands for water quality.

9.3.3 Increasing urban populations and decreasing rural populations

Figure 9.2 (PM, 2010) shows the formidable population increase that started in the 19th century and became consolidated in the 20th century in line with major health achievements.

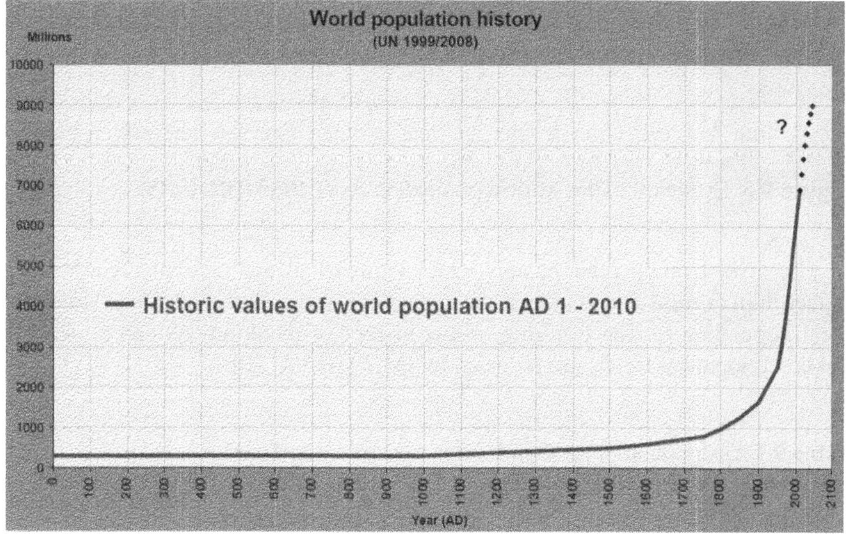

Figure 9.2 Evolution of the world's population in the last 20 centuries.

But the most significant increase has taken place in the last six decades, when the World's population has tripled. The population in 1950 was 2500 million, and today it stands at 7500 million (Figure 9.3). This is an imposing figure particularly when comparing it with the growth in the first 18 centuries, with a discrete growth of 600 million (from 400 million at the start of the first century to 1000 million at the start of the 19th century). Nobody knows what is in store. The United Nations (UN) contemplates different scenarios ranging from 9500 million to over 13,200 million as the highest value (UN, 2015).

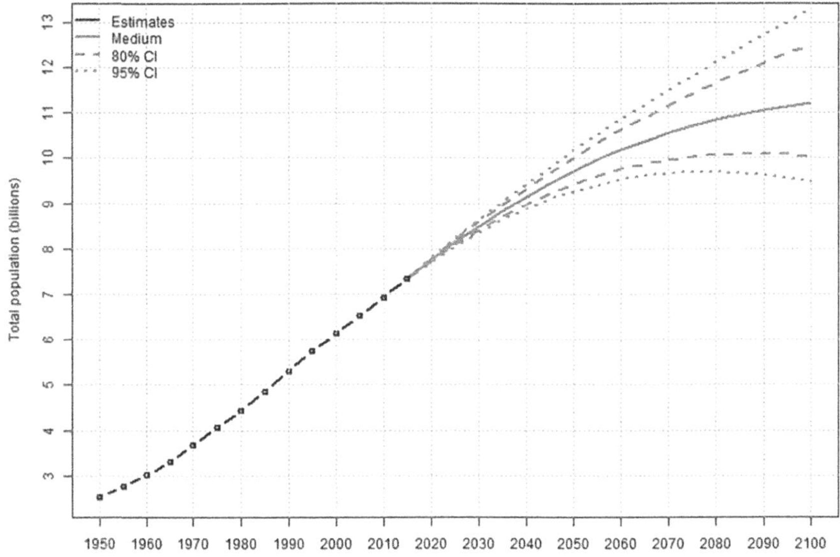

Figure 9.3 Evolution of the world's population as of 1950 (UN, 2015).

This growth on the other hand is very asymmetrical, and in urban areas is much higher than in rural areas. In fact, it was in 2007 when urban population exceeded rural population for the first time in the history of mankind. As can be seen in Table 9.1, identical trends are reported in Spain (BBVA, 2010).

Table 9.1 Distribution of rural and urban population in Spain.

Year	Total Population	Cities P ≥ 50,000	Cities 50,000 > P ≥ 1000	Rural Population (P < 1000)
1900	18,830,649	13.73%	74.16%	12.11%
2001	40,847,371	50.63%	45.53%	3.84%
2009	46,745,807	52.47%	44.31%	3.22%

(BBVA, 2010)

We are therefore witnessing unstoppable urban growth in detriment of rural growth. This means the challenge of sustainability in cities and towns is even bigger, without the necessary economy of scale. This is made worse by the fact that cities are undoubtedly growing faster than they should. In fact from the total of 8115 existing towns in Spain in 2015, 6799 (83.7%) had less than five thousand inhabitants and 4862 (60%) had less than one thousand inhabitants (INE, 2015b). A lot of imagination is therefore required to take the right approach to resolve the problems, and of course, to optimise the scarce economic resources to the maximum.

9.3.4 Growing urbanisation

The strong process of urbanisation in the territory, as a result of a growing population, notably alters the hydrological cycle of water. Before rainwater can filter through the earth and replenish the aquifers and nourish vegetation in the countryside (green water) once it has soaked into the ground, a large amount of it becomes run-off water relatively frequently flooding urban areas.

9.3.5 Climate change

The preceding challenge is of a more local nature, and largely depends on the circumstances involved in each case, but climate change, on the other hand, is a global effect which affects the entire world equally, although it does have different consequences. In the case of Spain, the forecasts (Milly *et al.* 2008) anticipate a severe reduction in the available resources and an increase, in frequency and intensity, of extreme phenomena (droughts and floods). Both facts mean that we will have to be far more efficient in our management of urban water. This is a challenge of extreme importance, that until now has been neglected. Today, after the Paris Summit, COP21, governments seem to be taking this formidable problem more seriously.

9.4 ANALYSIS, ONE CENTURY LATER, OF THE WATER – TOWN HALL BINOMIAL

After establishing the historical context of the urban water cycle and reviewing the formidable challenges future water management faces, we shall now analyse the current situation, and at the same time assess if municipal authorities are capable of tackling the huge challenges described on their own. The answer, as will be seen, is that in the majority of cases, they are not. The current problems are discussed which local corporations need to overcome on a daily basis, before discussing the direction they need to follow if all these obstacles are to be overcome in the next section. When doing so we should not forget that nearly 85% of towns in Spain have less than 5000 inhabitants. There will never be experts in these towns sufficiently qualified to make the most appropriate decisions. We shall now take a look at the problems to overcome.

9.4.1 Lack of professional capability by political decision makers

There is no argument about the fact that responsibility for urban water services should lie with the municipal authorities. In fact there is nothing to recommend this being any other way. But it is obvious that as time has gone by the complexity involved in sustainable management has increased significantly. A Mayor, or a newly elected Councillor is not really expected to be an expert in these problems. In any case, if there is interest in the subject, with enough common sense, and honesty in terms of putting the general interests of current residents, and above all future residents, before any short-term political interests, this will be sufficient. With good judgement, the most suitable decisions for the future can be identified. These parties should keep their eyes on the future and forget about the short-term interests that usually prevail in this world.

9.4.2 Lack of training by managers and technicians

There are three different levels of decision-making in the urban water cycle, which are in turn complementary to each other. The political level, which we have already seen, the managerial level and the technical level. The two latter respectively pertain to whoever controls the finances and the human resources, and whoever makes the technical decisions. In general, especially in medium- and small-size towns, training of these people is not at the level demanded by the growing complexity of these services, in terms of knowledge about urban water management. This is evident through examples of very bad practices (obviously from the perspective we are interested in in this document). For example, many towns have had a common treasury where the revenue from water is mixed with other municipal taxes. This is a rather inappropriate procedure because, among other problems, it prevents transparent management of the urban water cycle, or worse still, funds can be assigned to other items. Neither does it appear reasonable to charge other services, such as waste collection, in the water bill. Citizens should know the exact cost of each service. A third example of bad practice is that unfortunately far too many small towns do not even have plans of the water distribution network. The usual excuse that the plans are in the plumber's head is absolutely ridiculous. Under these circumstances it is very difficult to manage these services professionally.

9.4.3 Lack of environmental awareness by citizens

Mediterranean culture has bred citizens to believe that water is a public asset and therefore they have universal right to it at a cost of next to nothing. This conviction has been traditionally reinforced by the major infrastructures that have been developed, to a major extent subsidised by the public administration. Nobody has bothered to explain, regardless of it being appropriate to charge an environmental cost, that as a resource it is, and must be, free. But all the infrastructures required to take water to our taps, and those necessary to return it to the natural environment with suitable quality, entail a cost, which according to the Water Framework Directive (EU, 2000) users should pay for. Owing to its importance, this subject has already been discussed.

Because of their lack of environmental awareness and knowledge, citizens do not understand the need to raise water rates, which is a subject that is unfortunately debated at urban politics level when it should be a question of "State". The sustainability of a fundamental service for society at large is at stake. The wonderful Mediterranean culture concerning water has not adapted to the current problems, and worse still, sometimes politicians have even used it to their advantage, bolstering this lack of knowledge. The average citizen is intelligent and well-informed, and is able to identify demagogy and opportunistic discourse from honest approaches for the future. The current crisis has proven that we should not live beyond our means, since in the end all the bills have to be paid anyway. Removing water from the municipal political arena is a subject of vital importance, because if a politician did so, well-educated citizens would penalise it when realising the incorrect procedure.

Having emphasised the need to adapt training by all the players to the demands of new times, it is worth pointing out that this is not always the case. As mentioned earlier, there are always exceptions to confirm the rule. Furthermore, in the case of big companies, the economy of scale means they are able to have a highly qualified workforce. But in small towns, it is common, and only natural, for the situation to be the opposite if appropriate measures are not taken. This could mean regions sharing experts. The relevance of this question (lack of economy of scale) is dealt with in more detail later on in a specific section. In any case, the obstacles to overcome are of a more general nature, detailed as follows:

9.4.4 Atomisation of responsibilities

Sharing responsibilities makes global analyses and making the best decisions more difficult. In urban water affairs, whether directly or indirectly, there are too many departments with some kind of competence. Even though history explains the current situation (water supply is a service that permits very different viewpoints, each of them being a potential competence of the relevant administration department) the administration departments do need to adapt to the current context, which is a very complex affair due to the enormous inertia of the system itself.

Let us now take a look at the institutions (the list is different in each autonomous region, and is not complete), with some kind of responsibility in this service.

- The township, responsible for guaranteeing residents are provided this service. As said previously, this is only natural.

- The Hydrographic Confederation, administrator of the water resources that are used and controller of the condition of the environment where treated water is discharged.
- The Health Ministry, the ultimate party responsible for the quality of drinking water and manager of the National Water Consumption Information System (SINAC). We should not forget that these systems distribute the water that citizens drink (or should drink).
- The Regional Health Councils.
- The Provincial Councils in their role supporting the smaller villages and towns.
- The Price Commissions that approve the water rates in many communities.
- The Water Directorates in many regions.
- The relevant Ministry responsible for water, through the General Water Directorate.

It would be a good idea to rationalise the role of each institution as far as possible. The current, not particularly operational framework, is the result of the growing complexity of these systems and the lack of adaptation by the administration to current problems. For example, the problems concerning the quality of tap water (the legal demands increase over time, and in line with this the difficulties to meet them, because water sources are more and more polluted) justify active participation by the national Health Ministry and the corresponding Regional Councils.

9.4.5 Standards of service quality pending establishment

Whereas the quality of tap water has been established, the technical conditions for providing services are not regulated either at national or regional levels. The reasons explaining this situation are obvious. Non-potable tap water is a point of infection that can cause serious problems for public health, and therefore can generate social alarm. With that in mind, the central administration reacted diligently and legislated accordingly, as seen in the wide set of rules regulating this concern. The most relevant points are detailed as follows:

- *Ministry of the Presidency*: Royal DECREE 140/2003 of 7th February, establishing the health criteria for the quality of water for human consumption.
- *Ministry of Health and Consumer Affairs*: Development of Article 27.7 of Royal Decree 140/2003 of 7th February. Document drafted in consensus with the Regional Authorities and approved on 9th March 2005.
- *Ministry of Health and Consumer Affairs. Sub-directorate General of Environmental and Occupational Health*: Order SCO/1591/2005 of 30th May, pursuant to the National Drinking Water Information System (SINAC).
- *Ministry of Health and Consumer Affairs*: Order SCO/778/2009 of 17th March, pursuant to alternative methods for microbiological analysis of water for human consumption. Official State Journal of 31st March 2009.

- *Ministry of Health and Consumer Affairs. Secretariat General for Health. Directorate General for Public and Foreign Health Affairs. Sub-directorate General for Environmental and Occupational Health*: Substances for maintenance, cleaning and disinfection of surfaces in contact with water for human consumption. Madrid, 11th April 2011.

As can be seen, recommendations and rules have been issued concerning tap water, even clarifying the substances to use for cleaning and disinfecting surfaces in contact with water for human consumption. However, there is nothing set down about the technical conditions for providing the service, which include among others: specifying pressures, measurement, minimum flow rates to cover demand for water to extinguish fires etc. In other words, there are no service standards. In practice, some Town Halls with sufficient capacities, have their own service regulations which, to a higher or lower degree, do actually regulate these matters. By reading these documents the need to establish common criteria and adapt them to today's technological and social requirements becomes evident.

9.4.6 Confusing rules of play when outsourcing the service

The public/private debate concerning urban water is a subject that is intrinsic to the service. The water supply to the city of Madrid, the first in Spain, was established in 1858, and a few years later, in 1867, in Barcelona. Since then there have always been heated debates. In 1875 in Birmingham, the council bought the service from the private developers who started it because *"the water to be supplied to citizens is a subject of maximum interest for citizens and must be controlled by their representatives and not by private speculators"* (Thackray, 1990). Shortly afterwards, and not so far away (Amsterdam, 1898), the situation was repeated. The city bought the service from the people who created it, *since they were interested in extending the cover of the network. They were more concerned about the profits than providing a quality service* (Swemmer, 1990). More than a century later, heightened by the current economic crisis, the debate is still very much alive. In this debate those who defend public management always resort to the same old arguments.

One destiny, that of Birmingham and Amsterdam, was initially shared, but later split. The Dutch capital chose public management, as did nearly all of northern Europe. And it is worth mentioning that these countries, with Denmark at the head, pay the highest prices in the world. More than 7 euros per cubic metre – five times the average price in Spain. But obviously, when recovering all the costs, even the environmental costs, the service becomes sustainable. Nevertheless, Birmingham, as for all of England, is today the paradigm of private management. As soon as she came to power in 1979, Margaret Thatcher privatised service provision and the infrastructures. All in all, the usual procedure in administrations in debt who see privatising as a source of income to overcome hardships and deficits, and to guarantee compliance with a sacred principle all too often violated in Spain, that all the money from water is used to improve the service, regulating the industry.

The current economic situation is conducive to many Town Halls outsourcing management of these sources, believing this will be a source of external income. There is nothing wrong with this, providing things are done properly. Because if, as is all too often the case, the millions collected in water rates are assigned to other budgets, apart from violating the Water Framework Directive, the only other thing achieved is to mortgage off the service even more. Consequently, it would be advisable for a higher level institution to establish some minimum directives about the essential aspects that the Calls for Bids regulating contracts should contain before these services are outsourced. Among many other items, these should prevent revenue from water being assigned to other purposes. This could be one of the functions of the future regulator in Spain.

To end this subject, we must point out that ever since these services have existed, there have been numerous arguments in favour and against both types of management. And this will continue to be the case as can be seen in the recently created Public Water Network in Spain, which has rekindled the debate. It is obvious that both types of management have their advantages and disadvantages. In any case, the biggest disadvantages of public management are mainly due to inertia and lack of communication and coordination between the different administration departments. This is a situation that leads to a lot of easy demagogy, and we will therefore discuss it again in section 5. But, as said earlier, what matters is that decisions are made professionally and not so much the management model. Anybody who wishes to delve further into this subject can find some excellent analyses, both for and against in literature (Boland, 2007; OECD, 2009; Flecker *et al.* 2009; Pigeon *et al.* 2013). All in all, what really matters here, is for management to be professional, albeit public or private.

9.4.7 Prices and policy criteria

The lack of clear criteria when establishing the final price citizens have to pay for this service is probably the biggest weak point in the urban water cycle. This subject, owing to its relevance, has already been mentioned in previous pages. But at this point it is worth discussing the collateral damage that subsidies entail. The first of them is well known. The criteria for assigning subsidies to finance infrastructures, all too often is to please political criteria. Since all the costs are not recovered, most infrastructures are financed with funds from different sources (European, national or regional), and consequently, assigning these funds is often susceptible to political interests (the procedures are normally assigned if the party distributing the funds and who receives the funds belong to the same political party). This procedure is not advisable, as all too often the parties who receive most funds do not manage them well. This should not be the procedure assigning public funds which, as time goes by, will be in even greater shortage. In fact, in the period between 2014 and 2020 the European Union will not finance any new investments.

In some cases, *"ex ante"* criteria have been conducive to building a tank and increasing the pumping capacity from underground water with a deeper well, when what should have been done was to replace the pipes in an excessively leaky network. In others, work has been carried out which, because integration with the system was not planned into the design, is now useless. This has occurred over a number of years. One of the most striking cases is the Menorca desalination plant, built in the Ciudadela area. With production capacity of 10,000 m³/day, work was completed in 2011 but the pipes that should have connected the plant to the water distribution network for the cities where the water was required, were not built. And yet it has still been promised that water will be taken to consumers in 2016 (five years after completion of the plant), but it is still not clear how the pipeline to distribute water around the island will be financed. At least six years' life have been lost for the facility which in itself has a short life. This is a perfect example of how things should not be done.

9.4.8 Lack of transparency in urban water management

Water is a public asset that must be managed efficiently and transparently. This is particularly true in the case of urban water – the highest quality water. That said, today it is impossible for citizens to know the level of efficiency and competence involved in this vital service. There are several reasons to explain this. Firstly, because the politicians responsible for this issue are only concerned with water reaching citizens under reasonable conditions, the level of efficiency of water management is often considered a minor affair, particularly in small towns where it is managed by the Town Hall, and is unknown because nobody is bothered about controlling it. It only becomes a concern in periods of drought, or scheduled water cuts. When asked to save water, citizens question the coherence of it since the networks leak much more water than they are able to save.

Obviously in systems over a certain size, above all when management has been outsourced, the efficiency of the service is perfectly well known. But nobody demands this information (only the Health Ministry through SINAC collects data about water quality) which is to a certain extent public information. Furthermore, private operators believe it is sensitive, confidential information, because when renewing concessions this information could be of a certain advantage (efficiency is closely linked to the condition of the infrastructure, which is relevant information when planning mid- to long-term management), and are therefore reluctant to disclose it. On the other hand, the surveys promoted by the Spanish Supply and Sanitation Association (AEAS) which specifically ask about this matter, are incomplete (only AEAS members respond – a sample representing the big suppliers and the small, privately managed ones). They therefore only give a rough, optimistic idea of the situation. In private conversations with the same operators, they confess to doctoring the results.

9.5 THE PILLARS THAT SHOULD SUPPORT FUTURE URBAN WATER MANAGEMENT

Having seen the challenges facing urban water management in the 21st century, and the current weaknesses, this section outlines the strategies to follow in order to revert the situation. They are discussed as follows:

9.5.1 Improve training by decision-makers (at the three levels – political, managerial and technical) and raise environmental awareness in society at large

Change will not be possible without full involvement by all players. This must start with the general public, who must be aware of just what is at stake. Only with this full awareness will the public support unpopular short-term measures such as increasing water rates. But training policy-makers is equally important. Mayors must realise that these services will only be sustainable if decisions are made with a mid- to long-term outlook. They must therefore realise that water-related matters are strategic and must be taken out of the municipal political debate. This is the only way to enable design of strategic plans for the future. Finally, at managerial and technical level, decision-makers should be suitably qualified to ensure their decisions are the most appropriate for each specific case. In order to guarantee the best service at the lowest possible cost in all scheduled action, the cost – profit ratio must be maximised.

9.5.2 Diagnosis of the current situation

In order to design a suitable strategy it is important to know just what the starting point is. An analysis of the situation should include three blocks. The first is a technical diagnosis that establishes the levels of efficiency (particularly water efficiency) of the system and the quality at which the service is provided. The second is an economic analysis, comparing the current revenue to the amount that should be charged in order to recover all the costs (including amortisation of the system) to identify just how far away from economic sustainability the system is. The third is an analysis of the available human resources for adequate day to day management, and also assessing their ability to make the right decisions. This will obviously have to be done for each urban centre. These should be simple, basic analyses, but which are sufficiently representative to establish the real baseline for each urban water supply.

9.5.3 Designing strategic plans or the long-term commitment

Once the first two stages, just described, have been dealt with (although the first one should be ongoing, since raising awareness should be continuous), it is time to tackle the third stage. After identifying the shortcomings, we will be in a position to design strategic plans that should be accompanied with the relevant plans to finance them. Logically, gradual tariff rises over time should make it possible to execute these works. The order in which these items should be determined is obvious. First the technical side and afterwards the economic/financial side to permit it.

9.5.4 Follow-up and update of strategic plans

Strategic plans are established with a mid- to long-term outlook. But day to day incidents and action, or simply priority changes, can make it advisable to modify previously established plans. We should never forget the complexity of establishing tightly defined mid- to long-term plans, particularly in the case of urban water infrastructures in which circumstances and conditions change with time. This can all be corrected by means of suitable Heritage Infrastructure Management (HIM). As described, one of the main challenges water services face in this century is reverting the gradual ageing of the infrastructures. And since economic resources are scarce, the right decisions must be made at the right time, which is only possible through a good HIM.

9.5.5 Adopting measures to permit economic sustainability of rural communities

Within the global economic balance of an urban water service, the operating and maintenance costs (including personnel costs) are a very significant part. It would appear obvious that grouping together small rural communities would allow these costs to be shared, forming a relevant part of the balance statement, and therefore resolving some of the main problems these communities have to face. Only willingness and a political view of the future is required to do so. Therefore, trained decision-makers are essential. And if grouping these communities together is not sufficient to reach a level of sustainability without increasing water bills excessively, subsidies need to be resorted to ensuring they are conducive to efficiency and do not lead to squandering, as has been seen far too often. The United Nations estimates (UN, 2010) that water costs should not exceed 3% of a household's income, which is a similar figure to today's telephony costs (water, as we have seen, is 0.4% of the income per capita)

9.5.6 Establishing clear rules of play, whichever management method is used

The decision to outsource water supply services, when deemed appropriate, is one of the attributes that the Local Regime Regulatory Law confers on townships. We have already stated that the purpose of this work is not to discuss the specifics (advantages and disadvantages) of each management method, which are well known and have been debated in depth. Whichever option is taken, the main advantage of a private operator (know-how permitting optimisation of the available resources and committing to better value for money solutions) can be neutralised if the townships are properly organised and are backed by training from higher levels. A clear example of this can be seen in Holland where all water is under public management, sponsored by the State, and the companies have created a technology platform for support and to keep them updated in terms of technology. Since they are not competitors, they share developments and tools without any kind of reluctance. It is also worth mentioning that the most expensive water in the world is in northern Europe, with Denmark at the head, where water is also publicly managed. Hence, we are unable to associate public management with cheaper water rates. Logically, the high prices in Denmark are not due to inefficiency, but rather to recovery of all the costs, including environmental costs, whilst at the same time providing an excellent service.

Regardless of the management method finally chosen, when providing water services, the rules of play must be perfectly defined. Because fierce competition currently existing between different operators could lead us to believe the water business is extremely lucrative and that any price increase will serve to fill the operator's pockets even more. This is true to a certain extent ... but not quite. Because the formidable investments over recent decades (dozens of thousands of million euros) that have been spent on upgrading these systems have not been paid for by users, contrary to the Water Framework Directive. A large part of it has been financed from Brussels. The rest has been charged to the national and regional budgets. And that is how managers operate these services, only having to face the operating and maintenance costs (salaries, energy, etc.) and not much else. Current prices do not include renewal of the infrastructures taking water from where it is captured to users' taps, and then to where it is discharged. And that is how they manage to balance their accounts.

But the party wronged through this procedure is the citizen, whose urban water heritage is devalued (in Spain water infrastructures are always publicly owned, only management is outsourced). And in the new context of economic crisis, with the end of European funds for major investments, there will be no money to replace the infrastructures. Therefore, the principles of operating a service are very clear, starting off with recovering costs. Maintenance, and extension where necessary, must be guaranteed. Hence the Calls for Bids when outsourcing these services (if this is the chosen option) must be very well defined. Private capital evades uncertainty. The main difficulty attracting investors today, to help upgrade the urban water infrastructures and boost the economy, is caused by the political uncertainty with changes in municipal governance. Consequently, clear rules of play need to be established. In order to do so, and to achieve other objectives that will be mentioned later, a regulator is required.

9.6 TWO IMPORTANT FINAL NOTES

We are going through times of turbulence. Public administrations are in debt unemployment is over 20% in the working-age population, and basically millions of citizens are having a bad time of it all. This has all been conducive to the search for external resources for debt relief and to contribute to activating the economy. The administration in general, and Town Halls in particular, resort to privatisation as a solution to some of these problems, which is leading to strong social unrest especially when health services are privatised. It is therefore advisable to stress that outsourcing water management has nothing to do with privatising water or privatising health services.

This comment is made in view of the recent social initiatives (such as the Public Water Network) that have been very active defending water supply and sanitation as essential public services for all. They have been specifically asking Brussels for EU legislation to force governments to "guarantee and provide all citizens with appropriate drinking water and sanitation services". And they urge:

- Community Institutions and Member States to ensure that all citizens are given their right to water and sanitation.
- That water supply and water resource management should not be governed under "the domestic market rules" and water services should be excluded from liberalisation.
- The EU to double its efforts to achieve universal access to water and sanitation.

What they are asking for is obviously amazing. It is all reasonable, and nobody with some amount of common sense could disagree with it, with the exception of the tag *"water services should be excluded from liberalisation"*. This gives the impression that all urban water problems could be solved, as if by magic, just by preventing liberalisation of the service, ignoring major problems such as the fact that Town Halls, with a single Treasury, assign money from water to other purposes. This is equivalent to diverting levies from an outsourced service for other purposes, when what is really important is that these systems are managed efficiently, things are done properly and of course, water and sanitation are guaranteed as a human right in a system that is sustainable over time. And this, as mentioned earlier, can only be achieved by setting socially-aware prices, good management and recovering all costs. This is a far cry from outsourcing management or not.

I have nothing against outsourcing water services. Neither do I have anything against public management of these services, which citizens tend to be more sympathetic to. But we need to be clear about certain things, differentiating what is important from what is circumstantial. With water, as it is an indispensable resource for life, it is very easy to resort to demagogy and at the same time turn a blind eye to what is really important – good management and sustainability of these systems, at the lowest possible cost and with the highest respect for human rights and the natural environment. Let us now get to the point then, and not waste our energy on minor discussions.

The second note I referred to at the start of this section is that in these new times, subsidies and help have reached their end. It has been relatively easy to secure public money to finance new projects in the last two decades, without having to justify the need, or to prove that the work to be executed is the most suitable to resolve a specific problem without studying all the possible alternatives and then deciding on the best option in terms of value for money. As mentioned earlier, many tanks have been built at network headers that leaked extensively because there were subsidies for building tanks and not for reduce the leaking levels in the network. Money has been thrown down the drain, which a society in economic crisis can simply not afford. To apply for new subsidies (demanded by the European Union in the 2014–2020 six-year-plan) an *"ex ante"* analysis has to be duly justified (that the proposed remedy is the most suitable to resolve the problem). And once the action has been completed, it has to be proved that the established objectives have been achieved (*"ex post"* analysis). Basically the weak will always need help via subsidies (nearly always in rural communities). But the times when money was handed over without any kind of control have come to an end.

9.7 SERVICE REGULATION – THE SOLUTION TO A LARGE NUMBER OF PROBLEMS

There has to be a guarantee that all revenue collected from water is exclusively used to improve water management. To achieve this there has to be a control mechanism and this is one of the regulator's most important tasks. But there are many others too. The following points, because of their relation to what has already been discussed, summarise some of the tasks (aimed at resolving the identified weaknesses) that a regulator should perform. Depending on the profile, some will be more high key than others. These are basically things that today, despite the numerous different bodies with competences in the sector, nobody actually does, bringing to light the need to regulate the sector. Among others, they should:

- Establish service standards (except those that are already regulated such as water quality) in regulated water supplies.
- Promote compliance with the cost recovery principle, ensuring scheduled investments are actually made and that they meet the established goals.
- Support whoever needs it to define a tariff regime, complying with the cost recovery principle, and ensuring it is socially fair. And at the same time, take it outside the scope of local politics.
- Support and control service outsourcing processes, attending drafting of Calls for Bids regulating these contracts and then following-up on them.

- Provide technical support for towns that require it.
- Promote training at the three management levels (political, managerial and technical).
- Encourage the principle of solidarity between urban and rural areas. And, with a view to a better economy of scale, facilitate uniting services in rural areas.
- Promote transparency by encouraging competition between towns using "yardstick" regulation forcing publication of service quality indicators. This is the only way citizens will be able to know the level of competence of the people who manage their water cycle.

9.8 CONCLUSION

This paper has taken a look at the daunting challenges the urban water cycle is faced with. Growing difficulty just serves to prove that we cannot continue tackling these problems in the same that we have been until now. This is even more true in the current situation of economic crisis, with administration departments in huge debt and European funds for building new infrastructures in danger of extinction. And even though it is reasonable to believe the economic crisis will not last forever, it is no less obvious that things will never be the same. Subsidies are scarce, and in any case they need to be well justified. Institutions need to adapt to the new times, as does all human activity.

After discussing the challenges, we have analysed the weak points in the urban water – local corporation binomial. Afterwards, having identified the problems, we have set out the directives that should be followed in future urban water management. These changes are not easy, and will therefore not be implemented in a short period of time. They need to be implemented gradually (the inertia and enormous weight of the current water culture advise moving ahead cautiously), although firmly and decisively. And the figure of a regulator appears to be indispensable in this entire process. It has been seen that in many countries with similar problems, this huge challenge has already been tackled, although employing different strategies. Spain needs to do so as soon as possible.

9.9 REFERENCES

AWWA (American Water Works Asociation). (2012). *Buried No Longer: Confronting America's Water Infrastructure Challenge*. AWWA, Denver, Colorado, USA.
BBVA (Banco Bilbao Vizcaya Argentaria). (2010). *Población 51: La población en España: 1900–2009 Fundación BBVA*. BBVA, Madrid, España.
Boland J.J. (2007). The business of water. *Journal of Water Resources Planning and Management, ASCE*. May-June, 189–191.
Brundtland B.H. (1987). *Nuestro Futuro Común Editorial Alianza*. Madrid, España, 1987, 1989, 1992.

Copeland A. (2013). *Can a Small System Develop an Effective Asset Management Program?* Opflow January 2013, American Water Works Association, Denver, Colorado, USA.

El País diario. (2013). Evolución de la renta per cápita en las Comunidades Autónomas. 27 de Octubre de 2013 http://elpais.com/ Acceso el 4 de diciembre de 2015.

EPA (U.S. Environmental Protection Agency). (2013). Drinking Water Infrastructure Needs Survey and Assessment, Fifth Report to Congress Office of Ground Water and Drinking Water, Washington, D.C. 20460, EPA 816-R-13-006.

Flecker J., Hermann C., Verhoest K., Van Gyes G., Vael T., Vandekerckhove S., Jefferys S., Pond R., Kilicaslan Y., Cevat Tasiran A., Kozek W., Radzka B., Brandt T. and Schulten T. (2009). Privatisation of Public Services and the Impact on Quality, Employment and Productivity (PIQUE), Summary Report of the Project 'Privatisation of Public Services and the Impact on Quality, Employment and Productivity' (PIQUE). Vienna.

GFA (Green for All). (2011). *Water Works Rebuilding Infrastructure Creating Jobs Greening the Environment.* www.greenforall.org/resources/waterworks, Oakland, California, USA.

INE (Instituto Nacional de Estadística). (2015a). *Encuesta sobre el suministro de agua y el saneamiento.* Año 2013 Instituto Nacional de Estadística, Madrid, España.

INE (Instituto Nacional de Estadística). (2015b). *Relación de municipios y códigos por provincias de España.* Instituto Nacional de Estadística, Madrid, España.

Milly P.C.D., Betancourt J., Falkenmark M., Hirsch R.M., Kundzewicz Z.W., Lettenmaier D.P. and Stouffer R.J. (2008). Stationarity is dead: whither water management? *Science*, **319**(5863), 573–574. 1 February 2008.

NWAC (National Association of Water Companies). (2012). Economic Impact of Water and Wastewater Infrastructure Investment. NWAC: document_88e6f490-3e0e-4ac7-9736-608fafeba6d3.

OECD (Organization for Economic Co-operation and Development). (2009). *Private Sector Participation in Water Infrastructure.* Check List for public action Organization for Economic Co-operation and Development, Paris.

OECD (Organization for Economic Co-operation and Development). (2015). *OECD Principles on Water Governance Organization for Economic Co-Operation and Development.* Water Governance Programme, Paris.

Pigeon M., McDonald D.A., Hoedeman O. y Kishimoto S. (ed.) (2013). *Remunicipalización. El retorno del agua a manos públicas.* Transnational Institute, Amsterdam.

PM (Population Matters). (2010). Human Population History Report. http://www.population-matters.org/, accessed on December 4, 2015.

Swemmer F.F. (1990). Water supply and water resources management. In: Urban Water Infraestructure, K. Schilling and E. Porter (eds), Kluwer Academic Publishers, Dordrecht, The Netherlands. Páginas 173–188.

Thackray, J. (1990). Privatization of water services in the United Kingdom. In: Urban Water Infraestructure, 1st ed. K. Schilling and E. Porter (eds), Kluwer Academic Publishers, Dordrecht, The Netherlands.

UE (Unión Europea). (2000). Directiva 2000/60/CE del Parlamento Europeo y del Consejo de 23 de Octubre de 2000. Diario Oficial de las Comunidades Europeas, de 22.12.2000.

UN (United Nations). (2010). Human rights. *The Right to Water.* Fact Sheet, n° 35. Office of the High Commissioner for Human Rights. Geneva, Switzerland.

UN (United Nations). (2015). World population prospects. *Key Findings & Advance Tables.* 2015 Revision. United Nations, New York.

Viavattene C., Pardoe J., McCarthy S. and Green C. (2011). Cooperative Agreements between Water Supply Companies and Farmers in Dorset, Evaluating Economic Policy Instruments for Sustainable Water Management in Europe FP7 Environment.

Werkheiser I. and Piso Z. (2015). People work to sustain systems: a framework for understanding sustainability. *J. Water Resour. Plann. Manage.*, **141**(12), A4015002.

Chapter 10

Reasons to give grounds for urban water regulation in Spain

Félix Parra

Director general de Aqualia

10.1 STABILITY IN INVESTMENT FRAMEWORK FOR INFRASTRUCTURE

In investment and financing processes for comprehensive water cycle infrastructure, the different participating entities should have regulatory stability, which in turn provides legal and financial certainty, standardising and putting the following into objective terms:

- The structure of revenue and its stability throughout the management contract.
- The recovery of financial, capital, operational and investment costs (OPEX and CAPEX).
- Managing risks associated with demand and supply.
- Successful implementation of environmental legislation to obtain European standards (guarantee of infrastructure durability).
- The all-encompassing nature of the water cycle (conveyance, distribution, drainage and reuse).
- Mechanisms to establish and approve tariffs (local and regional levels).
- Subsidies available in the sector and compensation schemes.

Ultimately, *a stable framework that permanently addresses the issue of economic, social and environmental sustainability is required.*

The requirements of basic water supply are *continuity, quality, price and environmental protection*, and to strike a balanced approach – primarily in costs and service quality – it is necessary to reach a major consensus between the relevant actors, including citizens.

The current situation is Spain is completely decentralised, with the above-mentioned factors subject to local pseudo-regulation, within the scope of each municipality and Autonomous Community, but without state standards of service in place.

The water sector in Spain is "flawed" and must be effectively regulated. In this regard, we must bypass a political profile and steer towards the *professional sphere*; in other words, it should be a type of regulation that delivers *legal and financial stability* to meet the *social, economic and environmental goals* we are set by different European rules and legislations.

10.2 NEED FOR REGULATORY BODY

There are currently more than 8100 towns in Spain, equating to the same number of tariff structures and regulatory standards of water and sanitation services (protected under the Law Regulating the Basis of the Local Government). This leads to *significant disparity in the quality of service provided*, and *with regard to sustainability and efficiency* for each one of the current operators, whether public or private.

Some EU countries have established a national Regulatory Body to *oversee efforts to achieve management targets and to monitor the sustainability and efficiency of water and sanitation services*.

10.3 REGULATORY PERSPECTIVES

The following three perspectives must be accounted for when analysing the concept of regulation:

- "Institutional architecture" (commitments undertaken, credibility provided, regular and transparent accountability for the benefit of society in general, etc.).
- "Conduct" (standardisation of tariff policies, incentive system, quality and investment need analysis, etc.).
- "Structure" (horizontal concentration of operator's model, degree of vertical separation of generation and distribution activities, competition rules, etc.).

The incentive system may help shape the key variables. The conceptual framework of *"incentive-based regulation"* would establish financial mechanisms to control the quality of service provided by operators, which would prioritise *efficiency in management*.

10.4 INDICATORS

A Regulatory Body could not function without a simple yet fundamental management scheme *guided by management targets and indicators*. This is the only way to monitor the service provider and objectively assess the progress and result of its management.

Under this scheme, it is possible to gain a better understanding of the quality of service provided, allowing for comparison with quality standards, objectification of management, measurement of progress, identification of points for improvement, definition of continuous improvement projects, and comparison against models of excellence, etc.

10.5 AQUARATING AND BENCHMARKING

These two types of exercises, based on management indicators, aim to, with reference to the first one, offer a comprehensive and objective standard to assess performance and provide an overall rating for the service provider. The system thoroughly assesses each area for evaluation and the reliability of the information provided, as well as offering recommendations to improve management practice. To this end, it assesses eight different areas: Service Access, Service Quality, Operational Efficiency, Efficiency in Investment Planning and Implementation, Efficiency in Business Management, Financial Stability, Environmental Sustainability and Corporate Governance.

The second system aims to compare providers in order to facilitate a better understanding of quality in service provision. It essentially entails activities to:

* Compare management indicators (key performance indicators – KPIs) based on standard variables shared between different utilities.
* Identify Strong and Weak points as well as Opportunities, and Threats (SWOT analysis) in service provision, in economic, environmental and social variables.
* Define continuous improvement projects.
* Measure against models of excellence (within the scope of CSR, Collective Social Responsibility).

There is currently an international benchmarking process developed by the European Benchmarking Cooperation, where standard comparisons are carried out between different countries and companies in the sector.

10.6 STANDARDISATION OF REGULATIONS AND TARIFFS

The Executive Commission of the Spanish Federation of Municipalities and Provinces (FEMP) recently approved a *"model" Water Supply Regulation*, which was drawn up in collaboration with the Spanish Association of Water Supply and Sanitation (AEAS) (the model on sanitation is currently being finalised). It aimed to standardise the characteristics of the service delivered to customers (access rights, causes for cutting supply, implementation of meters, frequency of invoicing, payment terms and methods, supply installation plans, types of contracts, etc.).

FEMP and AEAS have also collaboratively drawn up a *Tariff Guide* with the aim of responding to needs emerging from the 2000 Water Framework Directive, the 2010 General Accounting Plan and the 2007–2015 National Water Quality Plan. The main aspects of the guide are as follows:

- The main aim is to achieve a *long-term financial-economic balance* of water and sanitation services.
- It intends to provide mechanisms and tools to enable the *total recovery of costs* (operational, environmental, financial, investment, capital and resource – or opportunity costs).
- It sets out several progressive tariff structures based on fixed and variable charges, facilitating transparency of cross-subsidies, standardising tariff analyses, and *recommending multi-year tariffs linked to management aims based on indicators and investment commitments*, as opposed to the current approval mechanisms (annually, with too much emphasis placed on social and political criteria).
- It stipulates the concept of "supply guarantee" in four levels (*quality guarantee, quantity guarantee, service guarantee and sustainability guarantee*).

10.7 REGULATION OF TARIFFS IN SPAIN AND POSSIBLE RESPONSIBILITIES

Analysing the suitability or lack thereof to establish a Regulatory Body of the Comprehensive Water Cycle Service should take account of two main areas for action:

- The capacity to analyse and approve tariffs and regulations to be applied by local governments and operators.
- The capacity to monitor compliance with the targets set out in the strategic management and service quality plans themselves.

Among other functions, this body could be responsible for:

- Conveying the policies established by Central and Regional Governments on the exploitation of water resources in tariff approvals.
- Enhancing transparency of the tariff for the end user (the Regulator must encourage transparency in information intended for citizens).
- Promoting the legal interests of customers.
- Acting as trained arbitrator between the local government and the service provider.
- Guaranteeing standardised tariffs and the possibility to compare services (benchmarking).
- Helping to quickly and effectively adjust tariff systems to respond to European Commission requirements (ensuring compliance with the deadlines set out under the Water Framework Directive).

- Auditing final charges for the financing of infrastructures.
- Strengthening the implementation of prices according to service quality indicators, i.e., correlating the price not only with the quantity delivered (m^3) but also with the service provided.
- Correlating efficiency in management with the service provided and the price thereof.

It is clear that there is a need to define the *main common and standardised variables* to identify costs, thereby enabling the cost and price comparison between the different regions and towns. As a result, *mechanisms for adjusting prices according to the efficiency* demonstrated with indicators would be set, linking them to compliance with investment commitments and the achievement of management goals (technical, economic and environmental parameters, etc.). *The more management costs are recovered, on the basis of measurable and reliable indicators, the better management will be for the end user.*

Insofar as internal organisation is concerned, the Regulatory Body should fulfil several basic conditions. It should:

- Be the *functioning authority* on regulated issues.
- Be *independent* from operators, politicians and financial and media circles.
- Have *administrative authority* to be able to apply and enforce the law and monitor its compliance.
- Be a *stable entity over time* with a clear intention to be permanent.
- Be in line with *European environmental and economic policy* applicable to the sector.

10.8 EXAMPLE OF REGULATION IN SIMILAR SECTORS: ENERGY

10.8.1 The spanish national energy commission (CNE)

The following list describes the progression of regulation in the energy sector leading to the founding of the CNE in 1998:

- "Stable Legal Framework" of 1988, which aimed to serve as a stable reference framework, primarily in the revenue system of companies that supplied electrical power, as well as in determining the electricity rate by applying cost optimisation criteria, something with which our sector's case draws noticeable parallels.
- "Protocol Establishing a Regulation for the National Electricity System" of 1996, signed by the Spanish Ministry of Industry and Energy and Spanish electrical companies.
- Law 54/1997 sparked a transformation of the electrical sector, as it deemed electrical supply essential for society to function. However, as opposed to former regulations, the new law rested on the conviction that guaranteeing

electrical supply, its quality and cost did not require state intervention but its own specific regulation; this is where the fundamental change lies in the perspective of the electrical sector. The important element to recognise is the abandonment of the notion of public service for the substitution of the notion of "guarantee of supply" and the incorporation of an indicative rather than decisive plan.

- The Spanish National Energy Commission (CNE, founded under Law 34/1998 of 7 October and subsequently developed by RD 1339/1999 of 31 July, which passed its regulation).
- The CNE protects the general interests of the electrical sector, and was established to encompass four key goals:
  - Supply guarantee
  - Supply quality and safety
  - Reasonable costs
  - Environmental protection

10.9 CONCLUSIONS

- It is healthy to publicly debate the issue of whether there is a need for a regulator and what this body should consist of. This debate should lay the foundations for its definition, after which it is necessary to work towards making sure it is established, responding to the interests of both sides.
- Dialogue between the different bodies involved is complicated yet essential; this discussion will lead to the truth and an agreement.
- Another important issue is deciding who controls the regulator, although it is necessary to eschew partisan political influence and the self-serving interests of companies.
- It is important to define the regulator's profile: technical, economic, political, … with independence to be earned through its actions and decisions. Another important aspect is knowing how much regulation costs and making sure it is self-sustainable in order to favour its independent status. This economic aspect may prove decisive when deciding on the body and the scope of the regulator (sole, autonomous, coordinated, etc.), without forgetting the necessary economy of scale.
- The regulator must play a role in sectoral legislation, in processes whereby concessions are awarded and in defining and monitoring the tariffs to be applied. All of the above must take place with transparent communication with the public and participation from all of the key players.

Ultimately, and as a general conclusion, this sector is clamouring for sole, independent and transparent regulation to professionalise the sector, removing it from short-term municipal political cycles. This would undoubtedly lead to enhanced legal and financial certainty, which would in turn enhance its credibility and provide information to and establish relationships with citizens, as exclusive recipients of more efficient water management.

10.10 REFERENCES

EFQM (European Foundation for Quality Management). The European Benchmarking Code of Conduct.

EUREAU (2011). Methodological guide on Tariffs, Taxes and Transfers in the European Water Sector.

European Commission (2003) *Green Paper on Services of a General Interest.* COM (2003) 270 final, 21 May 2003. [EU Commission – COM Document].

European Commission (2004) *White Paper on Services of General Interest. Communication from the Commission to the European Parliament, the Council, the European Economic and Social Committee and the Committee of the Regions.* COM (2004) 374 final, 12 May 2004. [EU Commission – COM Document]

FEMP y AEAS (2011). Guía de Tarifas de los Servicios de Abastecimiento y Saneamiento de Agua.

FEMP y AEAS (2011). Reglamento Tipo del Servicio de Abastecimiento de Agua Urbana.

IWA-BID (2014). AquaRating – A Rating System for Water and Sanitation Service Providers. [online] Aquarating.org. Available at: http://www.aquarating.org/en/

Waterbenchmark.org (2014). European Benchmarking Co-operation (EBC). [online] Available at: https://www.waterbenchmark.org/ (accessed 27 June 2016).

Chapter 11

International panel round table conclusions

Enrique Cabrera, Enrique Cabrera Jr. and*
Enrique Hernández Moreno[†]

*ITA, Universitat Politècnica de València

[†]Aqualia

The round table starts with a summary by the moderator of the main messages by each speaker which are collected here and validated by the speakers. Many of these messages were discussed afterwards at the round table session, and where relevant will be discussed in more detail later on. The first speaker was Jaime Melo (Portugal) who insisted on the need to properly integrated all the sides of this complex polyhedron comprising regulation of such a transversal service. The economic, environmental, health, social or technical aspects cannot be organised separately. Integration demands harmonising all points of view and mitigating the inevitable conflicts of interests between parties. He also insisted on the need to maintain permanent dialogue between operators and the regulator in Portugal. In other words, regulation not only consists of the inevitable "top-down" structure (what is legislated is to be implemented). For it to be effective and accepted, a "bottom-up" information flow from operators to regulator is also fundamental.

Michael Rouse (United Kingdom) emphasised two ideas. The regulator must be independent from political power and regulated companies. Politicians have the mission of legislating (defining and specifying all the themes to be regulated, such as the quality of the service), whereas the regulator is to implement approved legislation. Regulators must prove their independence from companies through objective reports defending the interests of users, who in the end are the ones who foot the costs of regulation. Without economic independence from political power, their neutrality will always be under threat.

Andrew Spears (Australia) explained the path that was followed to implement regulation in Australia, a continent/country consisting of autonomous states. The objectives and a general framework were defined at top level, for each State to develop in its own way. He emphasised the idea of independence of the regulator, providing some specific examples. In states where the regulator has been independent from political power, the results have been much better than where the regulator was susceptible to political directives. In particular, he quoted the Sidney desalination plant, built despite there being unanimity in the technical reports advising against it because it was not necessary. Time has proved the experts to be right, highlighting the folly of the political decision.

Jens Prisum (Denmark), as Director of a large company in the Danish sector, highlighted the principal message in his article. Danish regulation is only economic, and because of its complexity, only economists are able to understand it. It is like a black box that very few people understand. Critical of the model, he insisted that regulators should be very familiar with the sector (not the case in Denmark) and that is perhaps why dialogue between regulator and companies is lacking. Today this is only a "top-down" regulation.

Andrei Jouravlev (Chile), with 30 years' experience in the water industry in Latin America, said that in 70% of the countries there is some kind of regulation. And consequently there is a huge amount of variety in the models, all co-existing at the same time. From among the different models he ruled out local level regulation, because of its inoperability, which is rather surprisingly the current model in Spain. After an analysis of such a broad spectrum, he concluded by saying that the existence of a regulator is not a guarantee in itself. Common sense is required, along with capacity to act and political independence, the latter having been emphasised by all.

Wolf Merkel (Germany) explained the singular case of German self-regulation, also valid in neighbouring countries (such as Holland and Switzerland), and in view of the results, can boast having excellent urban water management. Perhaps because of that, nobody in Germany (users or companies) actually misses it. Technical regulation has been in force for many years through the sector association (DVGW) with health regulation by the relevant Federal Ministry (obviously in line with European Directives), whereas economic regulation is down to the federal states who establish specific directives to calculate and set tariffs. German culture and tradition mean that within this framework, the sector is in very good health. Merkel pointed out that this self-regulation model is very closely related to the country's culture, and in other frameworks it may not work well. This regulatory system needs to be acutely aware of the political and cultural system in the country where it is implemented.

Finally, María del Rosario Navia (International Development Bank, IDB) referred to the IDB and International Water Association (IWA) initiative to promote a system to assess urban water companies, leading to a tool called AquaRating which is described in her discussion. She emphasised the interest of this experience for any regulatory system, particularly during the period of establishing the metrics required to qualify how a company is working. She insisted greatly on the fact that nothing makes sense if the baseline data is not good quality. The cost is between thirty and fifty thousand dollars, required to pay the technical support when preparing reliable data and the fees of the auditors certified by AquaRating to certify the quality of the process. In exchange, the advantages for the company are clear – knowing their strengths and weaknesses.

11.1 THE DEBATE

After the speeches by the seven members, there was a round table lasting for nearly two hours with participation by the moderator (Enrique Cabrera), the panel of seven international experts and all the participants at the event (nearly 200). For the sake of briefness, the individual interventions are not detailed, but rather a synthesis of the subjects brought forward by the participants.

The first matter discussed in the debate was *"comprehensive regulation"*. As we have already seen, urban water is a transversal service with economic, environmental, health, social and service quality (technical) implications. Because of the scope of the problem, it is only natural for these responsibilities to be shared out between different institutions. Moreover, this is only natural because some of them (such as health control) are already covered in most cases, and it is not logical for newly created institutions to take on competences that are already perfectly well catered for. The role of a regulator should be to take on the new competences (those that time has proved are required or lacking) and those that experience has shown to be neglected to some extent. But nothing else. And the conclusion is obvious: effective regulation through one single institution is, as of a minimum size, operationally unviable.

Regulation of this activity through several institutions is also positive because in the case of conflicting interests, there can be debated dialogue between them. But it is also very important to properly define the objectives of each institution, in line with their objectives, and to define the tasks in order to avoid legal vacuums and overlapping competences. Promoting coordination and exchange of ideas between institutions, users and companies is fundamental. And this should all be carried out with utmost transparency. Concerning this matter, there is a Consultancy Council of 32 members in Portugal, with representatives of all the involved parties. This is necessary to improve the governability of the industry, with this subject being something several attendees showed interest in.

When talking about integration and the size of the sector to regulate, it was also mentioned that in some small countries the same regulator is in charge of all the public services (water, electricity, gas, waste and telephony), but this kind of regulation cannot assess the performance by companies in the sector at the right level of detail. This will always be "weak" regulation. That is the case in the United States.

Having established absolute unanimity about the need for *independent regulation*, particularly in the more sensitive aspects of regulation (obviously the economic aspects and prices), the second matter that was debated is as follows: Said in other words - Which is the most suitable model to ensure regulators are truly independent? And in addition to the economic independence of the regulator (paid by users), with regard to the need to separate control of the regulator from political cycles (always avoiding a new regulator with each new government), it was stated that there should be a clear definition of the profile the regulator should meet and the procedure to be followed to appoint it. These measures are very helpful to gain independence from political power.

Independence of companies is achieved with good work by the regulator. This must be proved by drafting reports with absolute transparency and freedom. This is the only way to earn the respect and trust of the sector, because if the regulator's job is not done properly, it will lose all credibility within a couple of years. Along these lines the idea was stressed that a good regulator needs to be an expert in the sector, since it is impossible to regulate something we do not know about. In response to the question about *who controls the regulator*, the answer was clear – it has to be the political power. And this is the case in England, where commissions created by Parliament perform this task.

Concerning the foregoing, the importance of *appointing the right regulator* was emphasised. A number of requirements regulators should meet were mentioned. After properly defining the professional profile of the post (which should include in-depth knowledge about the sector), a political appointment must be avoided at all costs. The suitability of establishing a maximum period for the post was also discussed, and that after leaving the post, the regulator should not be able to work in any other company in the sector until a certain time has elapsed (four years for example – a legislature).

Another widely debated subject was the *geographical scope* of action (national, regional or local) of the regulator, which is obviously an interesting issue for Spain. Since countries with very different regulatory models were represented on the panel (decentralised in Australia, homogeneous in England and Portugal), ideas flowed in abundance, proving there is not just one single answer to the question, but many aspects to consider. The first thing is that the regulation model must adapt to the current political structure, as stated previously. There is no point in swimming against the current. But it is worth having a common, integrating framework, the work of consensus between all the involved parties. This is the case in Australia where a national framework was created, on the basis of which different regional models were developed, with varying degrees of success, but in the end it was a model that reflected the political reality of the country. The second important issue is the cost of regulation, which should try to take advantage of the economy of scale as far as possible. In Portugal, for example, in the Azores where there is a significant level of political autonomy, they chose their own regulator. The result is that the cost per inhabitant is fifteen times higher than on mainland Portugal. There was absolute unanimity, and this has been emphasised a lot, that local regulation makes no sense at all.

The subject of the *cost of regulation* led to a heated debate. And the first reflection on this matter was that users, who in the end have to pay for regulation through their supply companies in the water bills, must know the real cost, and it must be a higher authority who actually controls that the cost is reasonable. The weight in the tariff logically depends on the economy of scale, which in any case should not be allowed to exceed a reasonable amount for the task of assessing company activity. For example, no more than a hundred companies to assess and the cost being positioned between 1% and 3% of the tariff paid by users, which was the general opinion about the total real cost of regulation, since it could go beyond the simple tariff cost, because the companies too will have to prepare a lot of information, and this will entail additional internal costs for them. And this was the first and only time when there were opposing opinions. On the one hand the English speaker argued that the data regulators compile are fundamental and therefore the company should be the most interested party in having this information, and it can therefore not be considered an additional cost. The Dane however, argued exactly the opposite. The economic regulation system implemented in Denmark is like a black box based on a model created by economists for economists. In this case the economic regulator is a state employee and does not entail an additional cost for users. But the companies need to hire auditors, economists and lawyers to help them meet the obligations imposed by the economic regulator. Basically the cost of regulation is not a trivial affair and is something to be taken very much into account when defining the regulation model, because obviously there is a direct relationship between the regulation model and the cost.

Another subject that led to a lively debate is how to convince the public about the need to commit to the mid-to-long-term rather than the current, traditional short-term outlook shared by politicians and users. In other words, how to persuade society about the need for tariffs to include capital costs and thus tackle the necessary issue of renewing the infrastructure. The response was unanimous: teach and inform through well-designed communication policies. Citizens are sensitive to environmental issues, so much so that they will be prepared to pay more for the services. But they should know why they are being asked to pay more. This reflection once again justified the need for greater participation by citizens in an absolutely transparent framework.

In relation to the need to regulate, one of the questions that arose from the audience particularly caught our attention. What sense does it make to regulate in a country such as Germany, which is not subject to direct control by a regulator, where the levels of water losses (6% on average) are some of the lowest? The answer was divided into three parts. The first was obvious. If this level of excellence has already been achieved, it is obvious that it is better not to change something that is already working well. The second, clarified by the German speaker, was that the published data (described in his paper) only include the figures from the big companies who, although they supply 70% or urban water in Germany, only represent a small percentage of the total number of companies in the industry. And that if the data were to include all the companies in the country, the figures would probably not be so good. Finally, there was a third reply by another member of the panel, that as the data is not audited by an independent regulator, the quality of the information has not been sufficiently verified, and is more than likely to err on the optimistic side.

Another concern of the audience was *how can a regulator*, convinced of the need to increase tariffs, *face up to political powers*, who are always reluctant to increase rates particularly during electoral times? The reply by the panel was categorical. One can only go against political power supported by reason, based on clear rules of play and a rigorous procedure for establishing tariffs, including the social side. And to achieve this companies must specify and itemise (according to the regulator's rules) all the costs on which the tariffs are calculated, which is not immediate. In Portugal for example, at the time regulation began (2003) more than 50% of the companies were incapable of determining their costs correctly. Today, one decade later, those calculations are part of their routine work. And we should not forget that the simpler and clearer this is, the better.

One of the most interesting questions was *how much profit is it reasonable for a regulator to accept when reviewing a company's accounts?* None of the speakers dared to put a specific figure on this amount. But they did suggest directives to establish it. The first is to associate this activity with another entailing a similar risk (always under the principle of higher risk, higher profit) and establish the profit in line with the laws of the market at the time. Since the risk involved in this activity is low, the profit should never be particularly significant. But obviously there does need to be a profit for the company to be sustainable, and it should also be sufficiently attractive to bring in new investment.

One of the attendees asked about *the time required to implement a regulation process*, and then about how long would it take from implementation before the first results could be seen. Obviously nobody wanted to give a specific figure, even when the English speaker said to what extent it was possible to answer the question. The transition period is the time that the companies take to gather all the data required by the regulator, which is not going to be immediate because many companies (as mentioned earlier) do not initially have this information.

It was also asked *who decides what to control and how to control it?* The reply by the panel was unanimous. Politicians have to decide what is necessary to control in order to guarantee the quality of the service when legislating. Afterwards, the regulator has to decide about the best way to "measure" the political objectives that have been established, and then in the third instance, through dialogue between the companies and the regulator, numerically establish what levels of quality are to be reached (specific objectives, for example the level of leaks). Over time, the established goals can be made more demanding, or can be relaxed where necessary. The regulation process should be dynamic. Concerning this matter, another idea related to the assessment and control process of companies, particularly when referring to benchmarking companies using predefined indicators, suggested the need to always compare "apples with apples", i.e., companies of a similar size. Along these lines Portugal defined nine "clusters" (groups of companies with similar characteristics), and companies are then compared to the other companies in the same cluster. It does not make much sense to compare a company that supplies millions of inhabitants with a company that just supplies a few thousand.

Another interesting subject that arose was *how can regulators keep their authority when faced with possible interference by other political bodies* that have more or less direct competences over urban water? Reference was made here to the competence commission and its ability to decide about tariffs authorised by the regulator. The response again was very clear. The regulator's mission is not to enter into debates with other state institutions, which are sometimes at a higher authority. In this case the reasonable course of events is to analyse the problem and issue the relevant report, so that another political authority, at an even higher level, can decide on the basis of the report.

Towards the end of the round table an attendee expressed in interest in the *role of the regulator in relation to the economic sustainability of urban water services in rural or small communities*. And sometimes it has no role, since in some countries (such as the case of Denmark) these services are outside the regulatory process. After a lively exchange of ideas it was concluded that regardless of whether or not they are included in the regulatory process, some measures of support need to be established. In Germany for example, since there is no regulator, it is the association (DVGW) that has taken responsibility for this subject. They have developed simple tools to (only an example) help calculate tariffs, they have defined specific indicators and standards of quality, they have encouraged joint management (the same expert simultaneously manages 10 rural systems), and finally, they have defined the profiles and training of the professionals who require these services. Moreover, these services are the subject of specific economic support by the regional governments. Basically, it is obvious that they need some kind of help that needs to be structured as reasonably as possible. This could be through a regulator or in any other way. But it is a reality that cannot be ignored if we do not wish to contribute to migration away from the countryside.

Finally, the last question put to the panel was divided into two parts. The first part referred to the facts and circumstances that had catalysed the process to create an urban water regulator in a certain country whereas the second part questioned what organisations or institutions had been most reluctant to participate in this process. And rather curiously, in all cases represented in the panel, the same answer was given. Implementation of a regulatory system cannot be an isolated action, but should be part of the general water policy. The process of controlling and managing urban water must be included in a broad legal reform. That was how it started in three countries (England, Australia and Portugal) where regulation has been implemented in recent decades. Furthermore, we should not forget that it is a long process. In Portugal for example, the process started back in 1993, when the figure of a regulator first appeared in the law. The first regulator was actually appointed in 2003, but another decade went by before the economic competences were transferred (this took place in 2014). Basically, regulation is not something that can be done overnight. In Australia the length of the process varied from state to state.

In reply to the second part of the question the two countries that replied (Portugal and England) were unanimous. It was the Town Halls who were most reluctant to participate in regulation. But, and it is important to emphasise this, as time has gone by they have become the strongest advocates of regulation, and in some cases (such as Portugal) they are pushing for competences to be reinforced as much as possible. They have seen that they have got rid of a problem in which they are not specialists, and better still, the daily service improves leading to increasing acceptance by citizens.

A very interesting round table event was brought to a close, with this last answer, where many and varied practical topics were covered. Most of them are not usually included in written papers. And that is precisely the added value of this final summary.

IWA Publishing's authorised EU representative for General Product
Safety Regulations is Diane D'Arras, 15 rue Duret, 75116 Paris,
France, e-mail: safety@iwap.co.uk.

Printed and bound by CPI Group (UK) Ltd, Croydon, CR0 4YY

12/05/2026

02108870-0003